Advanced Computational Fluid Dynamics for Emerging Engineering Processes - Eulerian vs. Lagrangian

Edited by Albert S. Kim

Published in London, United Kingdom

IntechOpen

Supporting open minds since 2005

Advanced Computational Fluid Dynamics for Emerging Engineering Processes - Eulerian vs. Lagrangian
http://dx.doi.org/10.5772/intechopen.77312
Edited by Albert S. Kim

Contributors
Santiago Lain, Baver Okutmuştur, Wahiba Yaïci, Evgueniy Entchev, Zhenhua Huang, Cheng-Hsien Lee, Wiesław Fiebig, Maciej Zawislak, César Augusto Aguirre, Carlos Sedano, Armando Benito Brizuela, Albert S. Kim

First published in London, United Kingdom, 2019 by IntechOpen
IntechOpen is the global imprint of INTECHOPEN LIMITED, registered in England and Wales, registration number: 11086078, 7th floor, 10 Lower Thames Street, London, EC3R 6AF, United Kingdom
Printed in Croatia

British Library Cataloguing-in-Publication Data
A catalogue record for this book is available from the British Library

Additional hard and PDF copies can be obtained from orders@intechopen.com

Advanced Computational Fluid Dynamics for Emerging Engineering Processes - Eulerian vs. Lagrangian
Edited by Albert S. Kim
p. cm.
Print ISBN 978-1-78984-372-9
Online ISBN 978-1-78985-031-4
eBook (PDF) ISBN 978-1-78985-032-1

For EU product safety concerns:
IN TECH d.o.o., Prolaz Marije Krucifikse Kozulić 3, 51000 Rijeka, Croatia, info@intechopen.com or visit our website at intechopen.com.

Meet the editor

Dr. Albert S. Kim is currently a full professor at the Department of Civil and Environmental Engineering at the University of Hawaii at Manoa. He has been teaching in the department from 2001, after he earned his MS (1997) and PhD (2000) in Civil and Environmental Engineering from the University of California at Los Angeles. Dr. Kim's scientific accomplishments include the National Science Foundation Faculty Early Career (CAREER) Award in 2005, the University of Hawaii Regents' Medal for Excellence in Research in 2006, and the University of Hawaii Regents' Medal for Excellence in Teaching in 2017. Professor Kim has published more than 60 peer-reviewed journal papers and six book chapters, and edited two open books.

Contents

Preface

Nature has four phases: solid, liquid, gas, and plasma. All phases except solid are in flowing states without having fixed molecular structures. In physics, matter and energy are known as exchangeable quantities, and water is a vital resource for human life as it makes up more than 70 percent of the human body. In engineering practices, humans create an artificial non-equilibrium state and induce nature to seek a local equilibrium within a reasonable time scale. Gradients of some physical quantities such as mass, heat, and momentum generate their fluxes that are amounts passed through a unit surface area per unit time. Engineering can be understood as a set of processes that convert these spontaneous fluxes into available resources. Fluid dynamics stems from Newton's second law for many individual particles interacting in viscous motion, which is represented using Navier–Stokes equations. Water is one of the representative fluids in engineering and sciences, having a pseudo-constant density, indifferent from temperature. A water molecule consists of one oxygen and two hydrogen molecules, of which molecular interactions determine the macroscopic properties. The hydraulic pressure can be explained, instead of force per unit area, as energy per unit volume that is equivalent to an energy density providing better understanding due to its scalar nature. Fluid mechanics is linked to statistical mechanics through pressure. The pressure is proportional to the negative gradient of the enthalpy per mass in an adiabatic system or Helmholtz energy per mass in an isothermal-isovolumetric system. Fluid dynamics is, by definition, a problem of solving the Navier–Stokes equation within a reasonable time frame. Computational fluid dynamics (CFD) is often used to analyze, optimize, and predict engineering phenomena and processes of practical interest. Transport of molecular matter such as salts, contaminants, reactants, and even macro-organics is often described using continuum equations that include convection, diffusion, reaction, and sourcing processes. When particles and solutes move relative to a moving fluid, (simultaneous) translation and rotation of multiple particles provide the intrinsically coupled feedback to the local fluid motion. Coupled simulations of fluid and particles are possible in principle but computationally challenging due to the mathematical complexity and computation demand. CFD simulation results can be, therefore, much more efficiently used if simulation runtime is significantly reduced so that more candidates of probable engineering scenarios are thoroughly investigated. On the other hand, highly accurate results are also of great necessity in fluid dynamics fundamentals. Open problems in CFD research literature include seamlessly merging fluid and particle dynamics while their relative motion is coupled due to the viscous characteristics of the solvent and rigorous analytic solutions for flow fields in geometrically less complex channels. In this vein, this book covers a wide range of state-of-the-art

CFD topics, providing future perspectives of advanced CFD methods as general tools of multi-physics simulations for sciences and engineering disciplines.

Albert S. Kim
Department of Civil and Environmental Engineering,
University of Hawaii at Manoa,
Honolulu, Hawaii

Section 1

Computational Particle Hydrodynamics

Chapter 1

A Coupling Algorithm of Computational Fluid and Particle Dynamics (CFPD)

Albert S. Kim and Hyeon-Ju Kim

Abstract

Computational fluid dynamics (CFD) and particle hydrodynamics (PHD) have been developed almost independently. CFD is classified into Eulerian and Lagrangian. The Eulerian approach observes fluid motion at specific locations in the space, and the Lagrangian approach looks at fluid motion where the observer follows an individual fluid parcel moving through space and time. In classical mechanics, particle dynamic simulations include molecular dynamics, Brownian dynamics, dissipated particle dynamics, Stokesian dynamics, and granular dynamics (often called discrete element method). Dissipative hydrodynamic method unifies these dynamic simulation algorithms and provides a general view of how to mimic particle motion in gas and liquid. Studies on an accurate and rigorous coupling of CFD and PHD are in literature still in a growing stage. This chapter shortly reviews the past development of CFD and PHD and proposes a general algorithm to couple the two dynamic simulations without losing theoretical rigor and numerical accuracy of the coupled simulation.

Keywords: computational fluid dynamics (CFD), computational fluid and particle dynamics (CFPD), dissipative hydrodynamics, tetrahedron mesh, mesh interpolation

1. Introduction

The first simulations of a liquid were conducted at the Los Alamos National Laboratory in the early 1950s, using the Los Alamos computer, Mathematical Analyzer, Numerical Integrator, and Computer (MANIAC). This computer was developed under the direction of Nicholas Metropolis, who is the pioneer of the modern (Metropolis) Monte Carlo simulation [1, 2]. The first MC simulation was conducted using the Lennard-Jones potential to investigate the material properties of liquid argon [3]. In the MC simulations, the phase space was searched to find more probable thermodynamic states and calculate macroscopic material properties (i.e., experimentally observable) using averages of the same physical quantity over the micro-thermodynamic states. In principle, the MC method assumes that the system of interest is in a static equilibrium state, and therefore, the time evolution is replaced by the phase-space averaging. The Boltzmann factor is used as a transition probability between two energy states.

The first CFD paper was published by Hess and Smith [4]. Their method is known as panel method as the surface was discretized with many panels. More accurate CFD work of advanced panel method can be found elsewhere [5, 6]. In general, CFD research is categorized by Eulerian and Lagrangian approaches, which are grid-dependent and meshfree, respectively. In the Eulerian category, the CFD performance regarding numerical accuracy and computational speed depends on how to discretize the computational domain. Popular methods include finite volume method (FVM) [7], finite element method (FEM), finite difference method (FDM), spectral element method (SEM), boundary element (BEM), and high-resolution discretization schemes.

The deterministic molecular dynamic (MD) method treats a particle as a point mass, which has a finite mass and other physical properties but does not have any volume and shape. Given external and interparticle forces, positions and velocities in linear motion are predicted using Newton's second law for N particles. When the motion of solute molecules in the solvent fluid is of concern, then the deterministic forces form solvent to solute molecules can be mathematically replaced by random fluctuating forces. These random forces have a zero mean over time, and its magnitude is determined by the dissipation fluctuation theorem [8].

Langevin's equation includes the hydrodynamic drag balanced by the random forces due to thermal fluctuation [9]. Solving N body Langevin equation is called Brownian dynamics (BD) [10]. Although BD can include effects of particle sizes in the dynamic simulations, it is fundamentally limited to the low concentration of solutes due to the Oseen tensor (see Appendix for details). Brownian dynamics (BD) was initially developed to reduce computational load by replacing deterministic interactions between a solute molecule and many solvent molecules by randomly fluctuating, probabilistic interaction as a net driving force. BD presumes a dilute solution, which means that a mean distance between solutes is much longer than the size of the molecule. A large number of solvent molecules exist around solutes, which is enough to exert random forces due to a tremendous number of collisions. When BD is applied to the dynamic motion of multiple particles, the lack of hydrodynamic interactions may provide erroneous results because the fluctuation-dissipation principle is not quantitatively well balanced. This limitation of BD to a single particle or a dilute solution is at the equivalent level of Stokes' drag coefficient, used to calculate the particle diffusivity.

Dissipative particle dynamics (DPD) is an updated version of BD, which specifically includes the interparticle hydrodynamic forces [11, 12]. Two functions, often denoted as w_R and w_D, are proposed to quantify the presumed relationship of pair-wise hydrodynamic forces/torques, determined by the dissipation fluctuation theorem. The proposed forms of the hydrodynamic interaction are vector wise as the real viscous forces are tensor-wise. Dissipative particle dynamics (DPD) updates BD by including an approximate form of tensor-wise hydrodynamic interactions as a pair-wise vector form. A force exerted on a particle has three types such as conservative, dissipative, and random forces, and DPD satisfies the fluctuation-dissipation theorem by balancing the dissipative and random forces.

Stokesian dynamics (SD) uses the grand mobility matrix to include the far-field hydrodynamic forces/torques [13]. In-depth comparative analyses of present particle dynamic methods can be found elsewhere [14]. This mobility matrix is inverted and updated by adding differences between near-field lubrication forces and far-field two-body interactions. As the lubrication force is proportional to the logarithm of the surface-to-surface distance between two particles, it diverges during events of particle collisions. Besides, if two particles are close to each other, i.e., the surface-to-surface distance is much smaller than the particle diameter, the SD formalism of near-field hydrodynamic contribution to the global hydrodynamic resistance (i.e., mobility inverted) often overestimates the real hydrodynamic repulsion forces. One of the primary reasons is that the lubrication theory assumes that particle surfaces are perfectly smooth, while in reality particle surfaces are rough and therefore surface friction also plays an essential role during collision events. Stokesian dynamics includes the most accurate estimation of the hydrodynamic tensors using the grand mobility matrix and pair-wise lubrication interactions. The fluctuation-dissipation theorem is better satisfied in SD by using many-body hydrodynamic interactions. In SD simulations, the fluid flow is given at any point in the computational domain as a combination of the unidirectional velocity, vortex velocity, and the rate of strain that is traceless and symmetric. The feedback interaction between a particle and fluid is already included in the expansion of Faxen rule. Due to the logarithmic characteristics of the lubrication forces, if two particles touch each other, they experience infinite repulsive forces. This characteristic of the lubrication force indicates the hard-sphere behavior of the colliding particles conceptually, implicitly assuming a perfectly elastic interparticle collision i.e., $\varepsilon = 1$ where ε is the coefficient of restitution.

Dissipative hydrodynamics (DHD) overcomes significant limitations of the particle dynamic method discussed above. DHD is a generalized method of which special cases converge to MD, BD, DPD, and SD by turning on or off specific force mechanisms. Details of DHD can be found elsewhere [14–16]. CFD and PHD were developed and applied without strong mutual influences. Particle tracking method can be viewed as a reasonable way to investigate the hydrodynamic motion of particles under the influence of ambient fluid flow. But, it has several fundamental limitations by neglecting particle density, particle shapes and sizes, and particle-fluid interactions. More importantly, the basic two-body interactions due to collisions, viscous flow, and electromagnetic properties are not included, and dispersion forces were dropped. In this light, the particle tracking method does not track particles but fluid elements moving along streamlines. While theories of particle hydrodynamics are not rigorously applied to engineering processes, this chapter includes a possible method to couple CFD and DHD in a seamless, robust way.

2. Dissipative hydrodynamics as unified particle dynamics

In this section, a unified view of preexisting particle dynamic method is discussed.

2.1 Overview

In the deterministic simulations, particle dynamics can be classified based on sizes of objects of interests. The purposes of particle dynamics are to provide the exact solution of a complex problem, bridging the theory and experiments. Note that the fundamental principles often provide governing equations, which were proven to be valid. It is difficult to solve a governing equation of a problem, if it has complex geometry and coupled boundary conditions. Experimental observations show natural processes using quantified information. Dissipative hydrodynamics is a generalized algorithm that unifies most of the preexisting particle dynamic simulation algorithms [16, 17].

At microscale of an order of nanometers, molecular dynamics deals with the motion of many molecules in various phases such as gas, liquid, and solid. Suppose that a system contains N molecules in volume V at temperature T. Newton's second laws of motion for this system is

$$m_j \boldsymbol{a}_j = m_j \frac{\mathrm{d}^2 \boldsymbol{r}_j}{\mathrm{d}t^2} = \sum_{i=1,\, i \neq j}^{N} \boldsymbol{F}_{ij} \tag{1}$$

where $\boldsymbol{F}_{ij}$ is a force exerted on particle j of mass m_j from particle i of mass m_i at time t. Position $\boldsymbol{r}_i$ and velocity $\boldsymbol{v}_i$ of molecule i are updated from time t to $t + \delta t$, i.e., from its initial values of $\boldsymbol{r}_i(t = 0) = \boldsymbol{r}_i^0$ and $\boldsymbol{v}_i(t = 0) = \boldsymbol{v}_i^0$, respectively, using the acceleration $\boldsymbol{a}_i$ determined using Eq. (1). Numerical evolution of Eq. (1) requires a specific algorithm for double integration [18–21].

A macroscale of order of millimeters, granular dynamics often called the discrete element method (DEM) includes specifically collision rules using the restitution and friction coefficients. During inelastic collisions of nonrotating particles, the particle kinetic energy is continuously lost, and their motion is decelerated. For a collision of rotating spheres, the surface friction provides an effective torque (as action and reaction) of the same magnitude and opposite directions to two colliding spheres. Rotational motion of a non-touching particle in a fluid medium generates angular dissipation of kinetic energy. Considering granules, i.e., non-Brownian particles, in a gas phase often neglect solute molecules and approximate the system as multiple particles undergoing the gravitational force field in a vacuum phase. As granular particle mass is much higher than that of colloids or nanoparticles in an aqueous solution, in a stationary phase the gravitational force is often balanced by normal forces developed at interfaces of particles to touching neighbors or a rigid wall. Implementation of the hydrodynamic lubrication interactions to granular particles is a difficult task, which requires an in-depth understanding of microscopic surface-deformation phenomena, linked to macroscopic particle motion.

A universal simulation method that can seamlessly include forces/torques exerted on arbitrary particles is therefore of great necessity. The method, first of all, should be able to:

1. Investigate the accelerating/decelerating motion of particles.

2. Satisfy the fluctuation-dissipation theorem for Brownian particles.

3. Include many-body hydrodynamic interactions.

4. Mimic inelastic collisions between spherical particles.

5. Apply constraint forces to form a nonspherical rigid body consisting of unequal spherical particles.

Dissipative hydrodynamics (DHD) has all the features required to be a universal simulation method for particle dynamics by taking specific advantages from MD, BD, SD, DPD, and DEM. A detailed review can be found elsewhere [14].

2.2 DHD formalism

Particle relaxation time. For a single particle motion in a viscous fluid, the governing equation can be in 1D space for simplicity:

$$m\ddot{x} = -\beta v + f'(t) \tag{2}$$

where m is the particle mass at position x, β is the drag coefficient, and f' is a random fluctuating force of zero mean:

$$\langle f'(t) \rangle = 0 \tag{3}$$

$$\langle f'(0) f'(t) \rangle = 2k_B T \beta \delta(t) \tag{4}$$

where T is the absolute temperature, k_B is Boltzmann's constant, and $\delta(t)$ is Dirac's delta function. Discarding the random force, one divides both sides of Eq. (2) by m to have

$$\frac{\mathrm{d}v}{\mathrm{d}t} = -\frac{v}{m/\beta} \tag{5}$$

which indicates that m/β has a dimension of time. This time scale is called the particle relaxation time, define as

$$\tau_p = \frac{m}{\beta} \tag{6}$$

which is a time scale that after a particle noticeably slows down after it starts moving with an initial velocity under the drag force. Stokes derived the drag coefficient

$$\beta = 6\pi\mu a_p \tag{7}$$

where μ is the absolute fluid viscosity and a_p is the radius of the primary particle [22].

Governing equation. A governing equation of DHD simulation is as follows:

$$\boldsymbol{M} \cdot \mathrm{d}\boldsymbol{v} = [\boldsymbol{Q}^p - \boldsymbol{R} \cdot (\boldsymbol{v} - \boldsymbol{U})]\mathrm{d}t + \boldsymbol{B} \cdot \mathrm{d}\boldsymbol{W} \tag{8}$$

where $\boldsymbol{M}$ is a diagonal matrix of mass and moment of inertia; $\boldsymbol{v}$ and $\boldsymbol{U}$ are translational/rotational velocities of the particle and the fluid, respectively; $\boldsymbol{Q}^p$ is the generalized interparticle and conservative force/torque vector; $\boldsymbol{R}$ is the grand resistance matrix; and $\boldsymbol{B}$ is the Brownian matrix of zero mean and finite variance:

$$\langle \boldsymbol{B} \rangle = 0 \text{ and } \langle \boldsymbol{B}^{\mathrm{tr}} \cdot \boldsymbol{B} \rangle = 2k_B T \boldsymbol{R} \tag{9}$$

where $\delta(t)$ is the Dirac-Delta function. And, $\mathrm{d}\boldsymbol{W}$ is the Ito-Wiener process [23, 24] having the following mathematical properties: $W_k = 0$ at $t = 0$, $W_k(t)$ is continuous, $dW_k\,(\equiv W_k(t+\delta t) - W_k(t))$ follows the normal distribution, and finally $\mathrm{d}\boldsymbol{W} \cdot \mathrm{d}\boldsymbol{W} = \mathrm{d}t$. The Brownian matrix $\boldsymbol{B}$ can be calculated by decomposing the grand resistance matrix such as

$$\boldsymbol{R} = \boldsymbol{A}^{\mathrm{tr}} \cdot \boldsymbol{I} \cdot \boldsymbol{A} \tag{10}$$

where $\boldsymbol{A}$ is the decomposed matrix to be obtained and $\boldsymbol{I}$ is the identity matrix. Statistically, the the identity matrix can be expressed as

$$\boldsymbol{I} = \langle \mathbf{C}^{\mathrm{tr}} \cdot \mathbf{C} \rangle \tag{11}$$

where $\mathbf{C}$ is a vector with zero mean and unit variance, i.e., $\langle C \rangle = 0$ and $\langle \mathbf{C}^2 \rangle = 1$, respectively. The Brownian matrix is defined as

$$B = \sqrt{2k_B T} C \cdot A \tag{12}$$

and substituted into (10) to provide

$$R = (A^{tr} \cdot C^{tr}) \cdot (C \cdot A) = (C \cdot A)^{tr} \cdot (C \cdot A) = \frac{B^{tr} \cdot B}{2k_B T} \tag{13}$$

Therefore, B is obtained by calculating a square root of R matrix, which is equal to A matrix of Eq. (10). The identity relationship of Eq. (11) is not satisfied at specific time t but statistically by taking an average of $C^{tr} \cdot C$ for a much longer period than the particle relaxation time τ_p. The effective force acting on a particle of a swarm of many particles in a viscous fluid is then represented as

$$Q^p - R \cdot (v - U) + \sqrt{2k_B T} C \cdot A \cdot W' \tag{14}$$

where $W' = dW/dt$. Although A is deterministically calculated to satisfy Eq. (10), $C^{tr} \cdot C = I$ is not valid at an instance but statistically. In the same sense, Eq. (11) is satisfied statistically.

Hydrodynamic tensors In Eq. (14), the generalized force requires a calculation of the grand resistance matrix R, which will allow to generate A. Consider particle i among N_p particles in a given volume V, translating with a linear velocity v_i and rotating with an angular velocity ω_i at an instantaneous position $r_i(t)$. In the absence of particles, the fluid flow at the center of particle i can be represented as $U^\infty(r_i)$. At a point $r = (x, y, z) \in S_i$ on surface S_i of particle i from the particle center r_i, the flow field is described as

$$V(r) = U^\infty(r_i) + \Omega^\infty \times (r - r_i) + E^\infty : (r - r_i) \tag{15}$$

where U^∞ is the unidirectional velocity, Ω^∞ is the vorticity,

$$\Omega^\infty = \frac{1}{2} \nabla \times V \tag{16}$$

and E^∞ is the rate of strain

$$E_{ij}^\infty = \frac{1}{2}\left(\frac{\partial V_i}{\partial x_j} + \frac{\partial V_j}{\partial x_i}\right) \tag{17}$$

which are evaluated at r_i. Because the rate-of-strain matrix is symmetric and traceless, the original nine components are reduced to

$$E_1^\infty = E_{xx}^\infty - E_{zz}^\infty = \frac{\partial V_x}{\partial x} - \frac{\partial V_z}{\partial z} \tag{18}$$

$$E_2^\infty = 2E_{xy}^\infty = \frac{\partial V_x}{\partial y} + \frac{\partial V_y}{\partial x} \tag{19}$$

$$E_3^\infty = 2E_{xz}^\infty = \frac{\partial V_x}{\partial z} + \frac{\partial V_z}{\partial x} \tag{20}$$

$$E_4^\infty = 2E_{yz}^\infty = \frac{\partial V_y}{\partial z} + \frac{\partial V_z}{\partial y} \tag{21}$$

$$E_5^\infty = E_{yy}^\infty - E_{zz}^\infty = \frac{\partial V_y}{\partial y} - \frac{\partial V_z}{\partial z} \tag{22}$$

The disturbance velocity field at the particle surface S_i is

$$\boldsymbol{v}_i^D(\boldsymbol{r}) = \boldsymbol{u}_i - \boldsymbol{U}^\infty + (\boldsymbol{\omega}_i - \boldsymbol{\Omega}^\infty) \times (\boldsymbol{r} - \boldsymbol{r}_i) - \boldsymbol{E}^\infty : (\boldsymbol{r} - \boldsymbol{r}_i), \quad \boldsymbol{r} \in S_i \tag{23}$$

where $\boldsymbol{u}_i$ and $\boldsymbol{\omega}_i$ are the translational and angular velocities of particle i, respectively. The translational/angular velocities and the rate of strain of particle i relative to the ambient flow field have then 11 degrees of freedom such as

$$\begin{aligned}(\boldsymbol{u}_i - \boldsymbol{U}^\infty, \boldsymbol{\omega}_i - \boldsymbol{\Omega}^\infty, -\boldsymbol{E}^\infty) &= \left(v_{ix} - V_x^\infty, v_{iy} - V_y^\infty, v_{iy} - V_y^\infty\right) \\ &+ \left(\omega_{ix} - \Omega_x^\infty, \omega_{iy} - \omega_y^\infty, \omega_{iz} - \Omega_z^\infty\right) \\ &+ \left(-E_1^\infty, -E_2^\infty, -E_3^\infty, -E_4^\infty, -E_5^\infty\right)\end{aligned} \tag{24}$$

For non-Brownian particles, the governing Eq. (8) is reduced back to that of Stokesian dynamics, which is Langevin's equation with the constant drag coefficient β, replaced by the grand resistant matrix $\boldsymbol{R}$.

When particle j is moving with linear and angular velocities of $\boldsymbol{u}_j$ and $\boldsymbol{\omega}_j$ under the influences of the ambient flow field characterized using $\boldsymbol{U}^\infty$, $\boldsymbol{\Omega}^\infty$, and $\boldsymbol{E}^\infty$, it experiences the hydrodynamic force $\boldsymbol{F}^H$ and torque $\boldsymbol{T}^H$. The stresslet $\boldsymbol{S}^H$ can be obtained but does not directly contribute to the particle acceleration. The generalized velocity and force are related through the grand mobility matrix μ^∞. Here, we use $\boldsymbol{q}$, $\dot{\boldsymbol{q}}$, and $\tilde{\boldsymbol{F}}$ for generalized coordinates, velocities, and forces, respectively:

$$\boldsymbol{q} = (\boldsymbol{r}, \boldsymbol{\theta}), \quad \boldsymbol{v} = (\boldsymbol{u}, \boldsymbol{\omega}), \quad \tilde{\boldsymbol{F}} = (\boldsymbol{F}, \boldsymbol{T})$$

The generalized relative velocity is

$$\Delta \boldsymbol{v} = (\boldsymbol{u} - \boldsymbol{U}^\infty, \boldsymbol{\omega} - \boldsymbol{\Omega}^\infty)$$

for both translational and angular motion. Then, the hydrodynamic interactions, i.e., forces and torques exerted on N bodies, can be expressed as

$$\begin{bmatrix} \Delta \boldsymbol{v}_j \\ -\boldsymbol{E}_j^\infty \end{bmatrix} = -\begin{bmatrix} \mu_{vF}^\infty & \mu_{vS}^\infty \\ \mu_{EF}^\infty & \mu_{ES}^\infty \end{bmatrix} \begin{bmatrix} \tilde{\boldsymbol{F}}_i^H \\ \boldsymbol{S}_i^H \end{bmatrix} \tag{25}$$

where $\boldsymbol{S}^H$ is the hydrodynamic stresslet. The matrix $\boldsymbol{\mu}^\infty$ (multiplied to $\left[\tilde{\boldsymbol{F}}_i^H, \boldsymbol{S}_i^H\right]^{\text{tr}}$) is called far-field grand mobility matrix. An inverse relationship of Eq. (25) is

$$\begin{bmatrix} \tilde{\boldsymbol{F}}_i^H \\ \boldsymbol{S}_i^H \end{bmatrix} = -\begin{bmatrix} R_{Fv}^\infty & R_{FE}^\infty \\ R_{Sv}^\infty & R_{SE}^\infty \end{bmatrix} \begin{bmatrix} \Delta \boldsymbol{v}_j \\ -\boldsymbol{E}_j^\infty \end{bmatrix} \tag{26}$$

where the matrix $\boldsymbol{R}^\infty$ (multiplied to $\left[\Delta \boldsymbol{v}_j, -\boldsymbol{E}_i^\infty\right]^{\text{tr}}$) is the far-field grand resistance matrix as an inverse of μ^∞, having the mathematical identity as

$$\begin{bmatrix} \mu_{vF}^\infty & \mu_{vS}^\infty \\ \mu_{EF}^\infty & \mu_{ES}^\infty \end{bmatrix} \cdot \begin{bmatrix} R_{Fv}^\infty & R_{FE}^\infty \\ R_{Sv}^\infty & R_{SE}^\infty \end{bmatrix} = \begin{bmatrix} I & 0 \\ 0 & I \end{bmatrix} \tag{27}$$

where I is the identity matrix. Note that $(\mu^\infty)^{-1} = R^\infty$ and $(R^\infty)^{-1} = \mu^\infty$ for grand matrices but $\left(R_{Fv}^\infty\right)^{-1} \neq \mu_{vF}^\infty$ for sub-matrices. The grand resistance matrix $\boldsymbol{R}$ in

Eq. (8) refers to R_{Fv}^{∞} of Eq. (26), and $R_{FE}^{\infty} \cdot E_j^{\infty}$ is an extra forcing term due to the rate of strain.

Note that at time t, the right-hand side of Eq. (25) is known, and one can calculate the generalized hydrodynamic force $\tilde{\boldsymbol{F}}_H$ by using an appropriate solver in numerical linear algebra [25, 26]. As noted in Eq. (15), the required information to evolve the motion of particle j, located at $\boldsymbol{r}_j$ with linear and angular velocities of $\boldsymbol{v}_j$ and $\boldsymbol{\omega}_i$, respectively, is the ambient flow field consisting of $(\boldsymbol{U}^{\infty}, ;\boldsymbol{\Omega}^{\infty}, -\boldsymbol{E}^{\infty})$ at the particle location. Unlike MD and BD, DHD explicitly includes the sizes of individual spherical particles. In this case, $\boldsymbol{r}_j$ indicates the position of the particle centroid or center of mass, and the fluid field is calculated at $\boldsymbol{r}_j$ without considering the presence of the particles. Therefore, the calculation of the ambient flow field is highly dependent on a mesh structure used for CFD simulations.

3. Computational fluid dynamics coupled with dissipative hydrodynamics

A general governing equation for fluid dynamics is Navier–Stokes equation, of which most general form for incompressible fluid is

$$\frac{\partial \boldsymbol{U}}{\partial t} + \boldsymbol{U} \cdot \nabla \boldsymbol{U} = -\frac{1}{\rho} \nabla P + \frac{\eta}{\rho} \nabla^2 \boldsymbol{U} \tag{28}$$

where $\boldsymbol{U}$ is the flow velocity, P is pressure, η and ρ are the viscosity and density of the fluid, respectively. In the adjacent space of particle j at $\boldsymbol{r}_j$, $\boldsymbol{U}$ can be expanded as expressed in Eq. (15). Most fluid dynamic problems have at least three boundaries, which are the inlet, outlet, and side walls. For inlet and outlet surfaces, Neuman and Dirichlet or mixed boundary conditions are often used to set values or conditions of $\boldsymbol{U}$ and P. On the wall, zero velocity and zero-gradient pressure are usually assigned. The former condition assumes that there are strong adhesion forces between solvent molecules and wall surfaces. The solvent molecules are fixed on the wall. The velocities of the wall-adsorbed solvent molecules are equal to those of the solid walls, which is zero for non-moving walls. Values of $(\boldsymbol{U}, P)$ are calculated at grid points in internal spaces surrounded by the boundary surfaces. The simulation accuracy and numerical convergence highly depend on the mesh structures. When CFD is coupled with particle dynamics, which is in this study, DHD, there are additional requirements that should be satisfied to evolve the motion of multi-particles moving in a fluid flow:

1. To identify a cell of the constructed mesh grid, which contains the kth particle's position, $\boldsymbol{r}_k$, for $k = 1 - N_p$

2. To calculate the distance between $\boldsymbol{r}_k$ and wall surfaces, if the particle is close to wall boundaries

3. To interpolate the flow field $(\boldsymbol{U}^{\infty}, ;\boldsymbol{\Omega}^{\infty}, -\boldsymbol{E}^{\infty})$ at $\boldsymbol{r}_k$ within a cell that contains kth particle

Possible methods to satisfy the three requirements are dependent on available CFD solvers and flexibility of applying customized modifications. Here we suggest fundamental approaches to meet the requirement numerically.

3.1 Geometric calculations

3.1.1 How to determine a cell containing the centroid position of my particle

Computational grid cells have various structures such as hexahedron, wedge, prism, pyramid, tetrahedron, and tetrahedron wedge. Among them, hexahedron followed by tetrahedron structures is widely used to generate mesh structure of bulk (internal) spaces. Cubic and rectangular shapes are representative structures of hexahedrons, consisting of eight vertices (points) and six rectangular (or square) surfaces. On the other hand, a tetrahedron cell has only four vertices (as compared to eight in hexahedron) and three triangular surfaces. Each of these two cell structures has its advantages and disadvantages in CFD simulations. Formation of hexahedron meshes is straightforward, and numerical solutions are well converged to provide accurate results within a tolerable error, especially if edge lines are well aligned to the flow directions. However, if a computational domain includes complex and curved surface structures such as human faces or globes, the hexahedron meshes often provide unrealistic small exuberances instead of well-curved surfaces. As three triangular surfaces surround the tetrahedron volume, it can form well-fitted boundary layers of arbitrary shapes. As edges of tetrahedrons cannot be fully aligned on a straight line, the numerical convergence of tetrahedron meshes is often more sensitive to the fineness of generated mesh structures than that of hexahedron meshes. To overcome this limitation, one can make tetrahedron meshes often finer than that of hexahedron meshes to solve the same problem with similar accuracy. Nevertheless, there are unique mathematical advantages of using tetrahedron meshes for coupled simulations of CFD and particle hydrodynamics.

Location test using volume calculation. A tetrahedron have four points, p_i, located at position $\boldsymbol{r}_i$ for $i = 1 - 4$, where $\boldsymbol{r}_i = (x_i, y_i, z_i)$. The volume of the tetrahedron can be calculated as

$$V = \frac{1}{6}\det(J) \tag{29}$$

where J is the Jacobian matrix given as

$$J(1,2,3,4) = \begin{bmatrix} 1 & 1 & 1 & 1 \\ x_1 & x_2 & x_3 & x_4 \\ y_1 & y_2 & y_3 & y_4 \\ z_1 & z_2 & z_3 & z_4 \end{bmatrix} \tag{30}$$

Using this mathematical relationship, one can easily test whether a particle position $\boldsymbol{r}_p = \left(x_p, y_p, z_p\right)$ is inside or outside the tetrahedron. If $\boldsymbol{r}_p$ is inside of the tetrahedron, then the total volume of the tetrahedron can be divided by four small pieces and their sum is equal to V:

$$V = V_p \tag{31}$$

using

$$V_p = \frac{1}{6}[J(p,2,3,4) + J(1,p,3,4) + J(1,2,p,4) + J(1,2,3,p)] \tag{32}$$

where, for example, $J(p,2,3,4)$ means the Jacobian matrix of Eq. (30) with the first column $(1, x_1, y_1, z_1)^{\text{tr}}$ replaced by $\left(1, x_p, y_p, z_p\right)^{\text{tr}}$. Conversely, if $\boldsymbol{r}_p$ is outside of the tetrahedron, we have an inequality of

$$V < V_p \tag{33}$$

If a particle is very close to one of the triangular surfaces of the tetrahedron, then the inequality check of Eq. (33) may not be done accurately. In this case, this location check can be extended to the nearest neighbor cells, especially one that shares the triangular surface that the particle is closely located.

For $\boldsymbol{r}_p$ within the tetrahedron, dividing both sides of Eq. (32) by V gives

$$1 = \xi_1 + \xi_2 + \xi_3 + \xi_4 \tag{34}$$

where, for $i = 1 - 4$,

$$\xi_i = \frac{J(\text{without } i)}{6V} \tag{35}$$

which is solely determined by the internal position $\boldsymbol{r}_p$. A relationship between $\boldsymbol{r}_p$ and $\boldsymbol{\xi}$ is

$$\begin{bmatrix} 1 \\ x_p \\ y_p \\ z_p \end{bmatrix} = \begin{bmatrix} 1 & 1 & 1 & 1 \\ x_1 & x_2 & x_3 & x_4 \\ y_1 & y_2 & y_3 & y_4 \\ z_1 & z_2 & z_3 & z_4 \end{bmatrix} \begin{bmatrix} \xi_1 \\ \xi_2 \\ \xi_3 \\ \xi_4 \end{bmatrix} \tag{36}$$

where the inverse of the 4×4 matrix can be analytically available. Eq. (36) assumes that $\boldsymbol{r}_p$ is located within the tetrahedron, but it can also use the particle location check as a better alternative to Eq. (33) because Eq. (36) requires one time solving of a 4×4 linear system, but Eq. (33) requires four times calculation of 4×4 Jacobian matrices.

3.1.2 How to calculate a distance between wall surfaces to the position of my particle

Using the four vertices of $(\boldsymbol{r}_1, \boldsymbol{r}_2, \boldsymbol{r}_3, \boldsymbol{r}_4)$ of a tetrahedron cell, one can identify four triangular surfaces of $S_1(\boldsymbol{r}_2, \boldsymbol{r}_3, \boldsymbol{r}_4)$, $S_2(\boldsymbol{r}_1, \boldsymbol{r}_3, \boldsymbol{r}_4)$, $S_3(\boldsymbol{r}_1, \boldsymbol{r}_2, \boldsymbol{r}_4)$, and $S_4(\boldsymbol{r}_1, \boldsymbol{r}_2, \boldsymbol{r}_3)$. For example, let us consider S_4 having vertices of $(\boldsymbol{r}_1, \boldsymbol{r}_2, \boldsymbol{r}_3)$. If we calculate relative position of vertices 2 and 3 with respect to vertex 1, denoted as

$$\boldsymbol{r}_{2/1} = \boldsymbol{r}_2 - \boldsymbol{r}_1 \tag{37}$$

$$\boldsymbol{r}_{3/1} = \boldsymbol{r}_3 - \boldsymbol{r}_1 \tag{38}$$

their cross product allows us to calculate a unit vector normal to surface S_4:

$$\boldsymbol{n}_4 = \frac{\boldsymbol{r}_{2/1} \times \boldsymbol{r}_{3/1}}{\left|\boldsymbol{r}_{2/1} \times \boldsymbol{r}_{3/1}\right|} \tag{39}$$

The particle position relative to $\boldsymbol{r}_1$ is $\boldsymbol{r}_{p/1} = \boldsymbol{r}_p - \boldsymbol{r}_1$. Then, the distance between surface S_4 and position $\boldsymbol{r}_p$ is simply

$$d = \left|\boldsymbol{n}_4 \cdot \boldsymbol{r}_p\right| \tag{40}$$

where the absolute value is necessary because $\boldsymbol{n}$ can direct inside or outside the tetrahedron volume, depending on choice of the reference position, $\boldsymbol{r}_1$. This wall-particle distance calculation is necessary when the surface S_4 is known as a wall boundary, so the distance d cannot be smaller than the particle radius a_p. After a single-step time evolution, if a particle is found overlapped with a wall surface, then collision rule will be applied to return the particle to the tetrahedron interior.

3.1.3 How to interpolate the flow field at the position of my particle

If a particle k is found inside a tetrahedron (of index l), the flow field consisting of $(\boldsymbol{U}^{\infty}, ;\boldsymbol{\Omega}^{\infty}, -\boldsymbol{E}^{\infty})$ needs to be interpolated using values calculated adjacent locations. A proper choice of a set of these locations are the vertex points of the containing cell. Based on the definition of $\boldsymbol{\Omega}^{\infty}$ and $\boldsymbol{E}^{\infty}$, the basic quantities needed are U_i^{∞} and $\partial_j U_i \equiv \partial U_i / \partial x_j$ for i and $j = 1 - 3$. (Note that pressure value is not required to calculate the generalized hydrodynamic forces $\tilde{\boldsymbol{F}}_H$.)

Four vertices of a cell can be represented as $\boldsymbol{r}_l = (x_l, y_l, z_l)$ for $l = 1 - 4$. A scalar quantity of interest S at each vertex point can be denoted as S_l. Then, the value of S at an arbitrary internal position $\boldsymbol{r}_p = \left(x_p, y_p, z_p\right)$ can be calculated as a linear superposition of S_l and ξ_l:

$$S_p = \boldsymbol{S} \cdot \boldsymbol{\xi} = \sum_{l=1}^{4} S_l \xi_l \tag{41}$$

For DHD simulations, S_p represents each element of the unidirectional velocity $\boldsymbol{U}^{\infty}$, vortex $\boldsymbol{\Omega}^{\infty}$, and the rate of strain $\boldsymbol{E}^{\infty}$, which are calculated using U_i^{∞} or its gradient $\partial_j U_i$.

3.2 Collision rules

Suppose there are two particles colliding each other, which are particle i and j located at $\boldsymbol{r}_i$ and $\boldsymbol{r}_j$, translating with $\boldsymbol{v}_i$ and $\boldsymbol{v}_j$ and rotating with $\boldsymbol{\omega}_i$ and $\boldsymbol{\omega}_j$, respectively. Relative position of particle i with respect to particle j is defined as $\boldsymbol{r}_{ij} = \boldsymbol{r}_i - \boldsymbol{r}_j$, and similarly the relative velocity of i to j is $\boldsymbol{v}_{ij} = \boldsymbol{v}_i - \boldsymbol{v}_j$. A normal vector from j to i is denoted as

$$\boldsymbol{n}_{ij} = \frac{\boldsymbol{r}_{ij}}{|\boldsymbol{r}_{ij}|} \tag{42}$$

After an instantaneous collision, the two particle have the following velocities [14]:

$$\boldsymbol{v}_i' = \boldsymbol{v}_i - \left(\frac{\mu_{ij}}{m_i}\right)\left[(1+\varepsilon^n)\boldsymbol{g}_{ij}^n + \frac{1-\varepsilon^t}{1+1/\tilde{J}\boldsymbol{g}_{ij}^t}\right] \tag{43}$$

$$\boldsymbol{v}_j' = \boldsymbol{v}_j + \left(\frac{\mu_{ij}}{m_j}\right)\left[(1+\varepsilon^n)\boldsymbol{g}_{ij}^n + \frac{1-\varepsilon^t}{1+1/\tilde{J}\boldsymbol{g}_{ij}^t}\right] \tag{44}$$

$$\boldsymbol{\omega}_i' = \boldsymbol{\omega}_i + \left(\frac{\mu_{ij}}{m_i a_i}\right)\left(\frac{1-\varepsilon^t}{1+\tilde{J}}\right)\boldsymbol{n}_{ij} \times \boldsymbol{g}_{ij}^t \tag{45}$$

$$\boldsymbol{\omega}_j' = \boldsymbol{\omega}_j + \left(\frac{\mu_{ij}}{m_j a_j}\right)\left(\frac{1-\varepsilon^t}{1+\tilde{J}}\right)\boldsymbol{n}_{ij} \times \boldsymbol{g}_{ij}^t \tag{46}$$

where $0 \le \varepsilon^n \le 1$ and $-1 \le \varepsilon^t \le 1$ are restitution coefficients in the normal and tangential directions, respectively, μ_{ij} is a reduced mass defined as

$$\mu_{ij} = \frac{m_i m_j}{m_i + m_j}, \tag{47}$$

and $\boldsymbol{g}_{ij}$ is the relative velocity at the point of contact defined as

$$\boldsymbol{g}_{ij} = \boldsymbol{v}_{ij} - \left(a_i \boldsymbol{\omega}_i + a_j \boldsymbol{\omega}_j\right) \times \boldsymbol{n}_{ij} \tag{48}$$

whose normal and tangential components are

$$\boldsymbol{g}_{ij}^n = \left(\boldsymbol{g}_{ij} \cdot \boldsymbol{n}_{ij}\right)\boldsymbol{n}_{ij} \tag{49}$$

$$\boldsymbol{g}_{ij}^t = -\boldsymbol{n}_{ij} \times \left(\boldsymbol{n}_{ij} \times \boldsymbol{g}_{ij}\right) \tag{50}$$

If a collision between a wall surface and particle i occurs, then the wall can be represented as a stationary spherical particle j having infinite mass and radius, i.e., $m_j \to \infty$, $v_j \to 0$, and $a_j \to \infty$:

$$\boldsymbol{v}_i' = \boldsymbol{v}_i - \left[(1 + \varepsilon^n)\boldsymbol{g}_{ij}^n + \left(\frac{1 - \varepsilon^t}{1 + \tilde{J}^{-1}}\right)\boldsymbol{g}_{ij}^t\right] \tag{51}$$

$$\boldsymbol{\omega}_i' = \boldsymbol{\omega}_i + \left(\frac{1 - \varepsilon^t}{1 + \tilde{J}}\right)\frac{\boldsymbol{n}_{ij} \times \boldsymbol{g}_{ij}^t}{a_i} \tag{52}$$

$$\boldsymbol{v}_j' = \boldsymbol{v}_j = 0 \tag{53}$$

$$\boldsymbol{\omega}_j' = \boldsymbol{\omega}_j = 0 \tag{54}$$

where

$$\boldsymbol{g}_{ij} \to \boldsymbol{v}_i - a_i\left(\boldsymbol{\omega}_i \times \boldsymbol{n}_{ij}\right) \tag{55}$$

$$\boldsymbol{n}_{ij} \to \boldsymbol{n} \text{ to particle } i \tag{56}$$

and $\boldsymbol{n}$ is the normal vector of the colliding wall surface inward to the liquid volume. During the collision of particle i with the wall, the wall is not moving so that $\omega_j' = \omega_j = 0$ and $\boldsymbol{v}_j' = \boldsymbol{v}_j = 0$.

4. Concluding remarks

Each of computational fluid dynamics and particle hydrodynamics is a challenging research topic, as applied to real engineering problems. Movements of particles (viewed as small solid pieces) in a moving fluid require rigorous interfaces to couple the two strongly correlated dynamic events. When the ambient flow pushes suspended particles in a liquid, dynamic responses of particles to exerting fluid change the fluid motion at the next time step, which returns to the particles with modified magnitude and direction. This fluid-particle (or fluid–solid) interaction is under the regime of Newton's third law, i.e., action and reaction. Multi-body simulations including the fluid-particle interaction are generally a difficult task. In this chapter, we briefly reviewed the CFD and PHD literature and discussed a feasible method to simulate the coupled fluid-particle motion within a reasonable time duration. A tetrahedron-based mesh is proposed to take the mathematical advantages of tetrahedron structure, consisting of four vertices and four triangular surfaces. Advantages of tetrahedron meshes include the following features: first, to test a location of particle within or outside a tetrahedron mesh-cell; second, to calculate a distance between a particle surface and a wall surface; third, to interpolate a fluid velocity and its gradient using those of values given at vertex locations; and finally to track each particle from one cell to the other by using a pre-built list of nearest neighbor cells. As DHD uses the SD algorithm for hydrodynamics of non-Brownian particles and Ito-Weiner process for random fluctuating forces, it can be used as a general particle hydrodynamic simulation method when it is coupled with CFD using specific mesh structures. Hexahedron-based meshes can be used for the same purpose with the intrinsic advantages of aligning grid edges to estimated streamline directions. When particles move in a channel of complex geometry, boundary surfaces can be better constructed using tetrahedron meshes. Open-sourced meshes include gmsh, tetgen, and netgen, which can import structure files, generate meshes, and export them to a CFD solver package. The current coupling algorithm of CFD and DHD is limited to cases that particle Reynolds number does not exceed 1.0, but this restriction can be avoided by considering dominant forces/torques exerted on particles and simulation time intervals as compared to the particle relaxation time. The new coupling method covered in this chapter may provide a new foundation in a coupled simulation of CFD and DHD including DEM.

Acknowledgements

This work was financially supported by the National R&D project of "Development of 1MW OTEC demonstration plant (4/5)" (PMS4080) funded by the Ministry of Oceans and Fisheries of the Republic of Korea.

A. Appendix

A.1 The Oseen tensor and Faxen law

The Navier–Stokes equation for a laminar flow is from Eq. (28):

$$-\nabla p + \eta \nabla^2 \boldsymbol{U} = -\boldsymbol{F}\delta(\boldsymbol{x}) \tag{57}$$

$$\nabla \cdot \boldsymbol{U} = 0 \tag{58}$$

where $\delta(\boldsymbol{r})$ is the Dirac-Delta function, which indicates

$$\delta(\boldsymbol{x} \neq 0) = 0 \tag{59}$$

$$\int_V -\boldsymbol{F}\delta(\boldsymbol{r})\mathrm{d}V' = -\boldsymbol{F} \tag{60}$$

where V' is a volume enclosing the origin $\boldsymbol{r} = 0$. The fundamental solution for $\boldsymbol{v}$ and p are

$$\boldsymbol{U}(\boldsymbol{x}) = \boldsymbol{F} \cdot \frac{\mathcal{G}(\boldsymbol{x})}{8\pi\eta} \tag{61}$$

$$p(\boldsymbol{x}) = \boldsymbol{F} \cdot \frac{\mathcal{P}(\boldsymbol{x})}{8\pi\eta} \tag{62}$$

The Oseen tensor $\mathcal{G}(\boldsymbol{x})$ for the fluid velocity is given by

$$\mathcal{G}_{ij} = \frac{1}{r}\delta_{ij} + \frac{1}{r^3}x_i x_j \tag{63}$$

which is independent of fluid properties. The Oseen tensor for the pressure p is

$$\mathcal{P}_j(\boldsymbol{x}) = 2\eta \frac{x_j}{r^3} + \mathcal{P}_j^{\infty} \tag{64}$$

where $\mathcal{P}_j^{\infty}$ is a constant at the ambient condition. The Faxen laws determine the hydrodynamic force and torque, especially, exerted on a sphere of radius a, moving with the linear and angular velocities of $\boldsymbol{u}$ and $\boldsymbol{\omega}$:

$$\boldsymbol{F} = 6\pi\eta a \left[\left(1 + \frac{a^2}{6} \nabla^2 \right) \boldsymbol{U}(\boldsymbol{x}) \right]_{\boldsymbol{x}=0} - 6\pi\eta a \boldsymbol{u} \tag{65}$$

$$\boldsymbol{T} = 8\pi\eta a^3 [\Omega(\boldsymbol{x}) - \boldsymbol{\omega}]_{\boldsymbol{x}=0} \tag{66}$$

which indicates that the Stokes flow requires a quadrupole $a^2 \boldsymbol{F} \nabla^2 \delta(\boldsymbol{x})$ in addition to a monopole of $-6\pi\eta a \boldsymbol{u}$, which is a drag on a sphere undergoing steady translation.

Author details

Albert S. Kim[1*] and Hyeon-Ju Kim[2]

1 Civil and Environmental Engineering, University of Hawai'i at Manoa, USA

2 Offshore Plant and Marine Energy Research Division, Korea Research Institute of Ships and Ocean Engineering, Korea

*Address all correspondence to: albertsk@hawaii.edu

References

[1] Metropolis N, Ulam S. The Monte Carlo method. Journal of the American Statistical Association. 1949;**44**(247): 335-341

[2] Metropolis N, Rosenbluth AW, Rosenbluth MN, Teller AH, Teller E. Equation of state calculations by fast computing machines. The Journal of Chemical Physics. 1953;**21**(6):1087-1092

[3] Wood WW, Parker FR. Monte carlo equation of state of molecules interacting with the lennard-jones potential. i. a supercritical isotherm at about twice the critical temperature. The Journal of Chemical Physics. 1957; **27**(3):720-733

[4] Hess J, Smith A. Calculation of potential flow about arbitrary bodies. Progress in Aerospace Sciences. 1967;**8**: 1-138

[5] Rubbert P, Saaris G. Review and evaluation of a three-dimensional lifting potential flow computational method for arbitrary configurations. In: 10th Aerospace Sciences Meeting. Reston, VA: American Institute of Aeronautics and Astronautics; 1972

[6] Maskew B. Prediction of subsonic aerodynamic characteristics: A case for low-order panel methods. Journal of Aircraft. 1982;**19**(2):157-163

[7] Hyman JM, Knapp RJ, Scovel JC. High order finite volume approximations of differential operators on nonuniform grids. Physica D: Nonlinear Phenomena. 1992;**60**(1–4):112-138

[8] Kubo R. The fluctuation-dissipation theorem. Reports on Progress in Physics. 1966;**29**:255-284

[9] Langevin P. Sur la the'orie du mouvement brownien. Comptes Rendus de l'Académie des Sciences (Paris). 1908;**146**:530-533

[10] Ermak DL, McCammon JA. Brownian dynamics with hydrodynamic interactions. The Journal of Chemical Physics. 1978;**69**(4):1352-1360

[11] Hoogerbrugge P, Koelman J. Simulating microscopic hydrodynamic phenomena with dissipative particle dynamics. Europhysics Letters. 1992;**19**: 155

[12] Koelman J, Hoogerbrugge P. Dynamic simulations of hard-sphere suspensions under steady shear. Europhysics Letters. 1993;**21**:363

[13] Brady JF, Bossis G. Stokesian dynamics. Annual Review of Fluid Mechanics. 1988;**20**:111-157

[14] Kim AS, Kim H-J. dissipative dynamics of granular materials. In: Granular Materials. Rijeka: InTechOpen; 2017. pp. 9-42

[15] Kim AS. Constraint dissipative hydrodynamics (HydroRattle) algorithm for aggregate dynamics. Chemistry Letters. 2012;**41**(10): 1285-1287

[16] Kim AS. Dissipative hydrodynamics of rigid spherical particles. Chemistry Letters. 2012;**41**(10):1128-1130

[17] Kim AS, Kang S-T. Microhydrodynamics simulation of single-collector granular filtration. Chemistry Letters. 2012;**41**(10): 1288-1290

[18] Verlet L. Computer 'experiments' on classical fluids. I. Thermodynamical properties of Lennard-Jones molecules. Physics Review. 1967;**159**(1):98-103

[19] Martys NS, Mountain RD. Velocity Verlet algorithm for dissipative-particle-dynamics-based models of suspensions. Physical Review E. 1999; **59**(3):3733 LP-3733736

[20] Hockney RW, Eastwood JW. Computer Simulation Using Particles. New York: Adam Hilger; 1988

[21] Hockney RW. The potential calculation and some applications. In: Methods in Computational Physics. Vol. 9. New York: Academic Press; 1970

[22] Stokes GG. On the effect of internal friction of fluids on the motion of pendulums. Transactions of the Cambridge Philosophical Society. 1851;**9**:1-106

[23] Wiener N. Differential space. Journal of Mathematical Physics. 1923;**58**:31-174

[24] Ito M. An extension of nonlinear evolution equations of the K-dV (mK-dV) type to higher orders. Journal of the Physical Society of Japan. 1980;**49**(2): 771-778

[25] Anderson E. LAPACK Users' Guide. Philadelphia, PA: Society for Industrial and Applied Mathematics; 1987

[26] Blackford LS, Choi J, Cleary A, D'Azevedo E, Demmel J, Chillon I, et al. ScaLAPACK User's Guide. Philadelphia, PA: Society for Industrial and Applied Mathematics; 1997

Chapter 2

Response Behavior of Nonspherical Particles in Homogeneous Isotropic Turbulent Flows

Santiago Laín

Abstract

In this study, the responsiveness of nonspherical particles, specifically ellipsoids and cylinders, in homogeneous and isotropic turbulence is investigated through kinematic simulations of the fluid velocity field. Particle tracking in such flow field includes not only the translational and rotational components but also the orientation through the Euler angles and parameters. Correlations for the flow coefficients, forces and torques, of the nonspherical particles in the range of intermediate Reynolds number are obtained from the literature. The Lagrangian time autocorrelation function, the translational and rotational particle response, and preferential orientation of the nonspherical particles in the turbulent flow are studied as function of their shape and inertia. As a result, particle autocorrelation functions, translational and rotational, decrease with aspect ratio, and particle linear root mean square velocity increases with aspect ratio, while rotational root mean square velocity first increases, reaches a maximum around aspect ratio 2, and then decreases again. Finally, cylinders do not present any preferential orientation in homogeneous isotropic turbulence, but ellipsoids do, resulting in preferred orientations that maximize the cross section exposed to the flow.

Keywords: kinematic simulations, Lagrangian tracking, nonspherical particles, response behavior, preferential orientation

1. Introduction

Nowadays, the use of numerical simulation techniques to assist the development and optimization of industrial processes dealing with turbulent multiphase flow has been included as one more step in their layout. Examples of them include pneumatic conveying, fluidized bed reactors, cyclones, classifiers, or flow mixers. Industrial sectors where such processes are important are the chemical, food, or paper industries as well as electric energy production. Due to the complexity of the involved flow, a great majority of simulations are carried out under Reynolds-averaged Navier-Stokes (RANS) in connection with an appropriate turbulence model to describe the turbulent dynamics of the carrier phase.

Two main frames are employed for the description of complex multiphase flows: the two-fluid model or Euler-Euler and the discrete particle models or Euler-Lagrange. In both of them, particles are approximated as point masses being transported in the carrier phase flow field; the solution of the flow around individual elements is usually too expensive and cannot be afforded. In the two-fluid model, both phases are conceived as two interpenetrating continua [1] whose properties are described by sets of partial differential equations. In the Euler-Lagrange approach, the discrete elements are considered as individual objects whose dynamics is governed by a Lagrangian motion equation. Therefore, to obtain the discrete phase variables in the computational domain, a large enough number of discrete element trajectories must be computed. On each particle, appropriate forces act reflecting the various microprocesses taking place at the element scale such as fluid-particle turbulent interaction, particle-(rough) wall interactions, and interparticle collisions [2]. Such technique is especially appropriate for the description of disperse multiphase flow, where usually particles have a size distribution, in confined domains where particle-wall collisions play a predominant role as pneumatic conveying, separation, and classification processes. Both techniques, two-fluid model and Euler-Lagrange, have been applied mainly considering spherical particles. This means that the forces due to the flow (drag and lift) as well as the microprocesses modeling, wall-particle and inter-particle interactions, are assumed to be for spherical-shaped elements [3]. In practical situations, however, nonspherical particles are encountered, either of irregular shape, either with well-defined shapes (fibers or granulates). For example, the paper industry uses large amounts of turbulent liquid to handle and transport the fibers that compose the paper pulp. Besides, such particles in the flow experience particle Reynolds numbers larger than one, $Re > 1$. For such particles, the most relevant transport mechanisms such as aerodynamic transport, wall-particle interactions, and interparticle collisions are substantially different than those for spherical particles.

With the objective of performing the numerical simulation of turbulent flows laden with nonspherical particles, additional information about the forces and torques due to flow (drag and lift forces and pitching and rotational torques due to the shear flow and particle rotation) is needed. It is known that for regular nonspherical particles, that is, ellipsoids or fibers, such forces depend on particle orientation with respect to the flow. For instance, fiber orientation plays a major role in chemical processes as injection, compression molding, or extrusion in which the mechanical properties of the suspensions are determined by the orientation distribution.

For the Stokes regime, particle Reynolds number much lower than 1, the behavior of the nonspherical particles can be determined by analytical methods. Forces and torques acting on an ellipsoidal particle were analytically computed by Jeffery [4]. In a series of papers, Brenner determined the forces due to the flow acting on arbitrary-shaped nonspherical particles in the Stokes regime under different flow configuration by means of theoretical methods [5]. In the creeping flow regime, also with particle Reynolds number much lower than 1, Bläser [6] computed the forces acting of the surface on an ellipsoid in free motion for different flow situations, which allow him to suggest a simple criterion for particle breakup.

The drag coefficient for particle Reynolds number higher than 1 must be obtained by experiments, physical or numerical, as the analytical methods are not applicable any more.

The experimental studies to determine the drag coefficients for nonspherical particles employ wind tunnels or sedimentation vessels. For a moderately wide particle Reynolds number range, there exist results for thin discs [7], isometric irregular particles [8], cylinders and plates [9], discs [10], and discs and cylinders [11]. Drag coefficients were developed in all cases only for certain particle orientations. Compiling such results, different correlations have been developed in terms of particle shape [12–14]. As representative parameter of the particle shape, two options appear to be dominant: the spherical particle equivalent diameter d_p, and the sphericity, defined as the ratio between the surface of the spherical particle equivalent diameter and the actual surface of the nonspherical particle. None of such correlations takes into account the dependence of the drag coefficient with the particle orientation in the flow. There exist some correlations that consider such dependence [15–17] and, therefore, they are appropriated to be implemented in a Lagrangian computation scheme. Nevertheless, corresponding results for lift and pitching torque coefficients are still not equally available. One of the first results for the different coefficients in terms of the orientation of elliptic particles was obtained by Hölzer and Sommerfeld [18] using the lattice Boltzmann method (LBM). Vakil and Green [19] used DNS to study the drag and lift coefficients of cylindrical particles depending on their orientation and aspect ratio, for Reynolds number up to 40, providing a correlation for them. In such work, the underlying flow field was assumed to be uniform and the flow around the particle was completely resolved. More recently, Zastawny et al. [20] applied DNS in the frame of the implicit mirroring immersed boundary (MIB) method to obtain the flow coefficients for four different ellipsoids. The authors provide specific correlations for the drag, lift, pitching torque, and rotational torque coefficients depending on the particle Reynolds number and orientation but without including the effect of the aspect ratio. The covered Reynolds number range was up to 300. Ouchene et al. [21] determined with DNS the drag, lift, and pitching torque coefficients for prolate ellipsoids with aspect ratio up to 32 and adjusted their results to proper correlations that include the effects of particle orientation and aspect ratio up to a Reynolds number of 300.

The first numerical computations of very small nonspherical particles in pseudoturbulent flow were performed by Fan and Ahmadi [22] and Olson [23]. The hydrodynamic forces and torques were computed by the theoretical coefficients of the Stokes regime. Olson [23] estimated the time step for the translation and rotation motions in function of the fiber length, obtaining the corresponding dispersion coefficients. Fan and Ahmadi [22] showed that the dispersion of both, translation and rotation, was reduced with the fiber length. However, Olson [23] found a different result in the case of ellipsoidal particles. Lin et al. [24] investigated numerically the distribution of the orientation of the fibers in a developing mixing layer, comparing the obtained results with experiments. The fiber length was smaller than the Kolmogorov scale, so they employed the forces due to the flow of the Stokes regime. Zhang et al. [25], Mortensen et al. [26], and Marchioli et al. [27] studied the transport and deposition of ellipsoidal particles in a turbulent channel flow using direct numerical simulation (DNS). Again the hydrodynamic forces and torques were computed with Stokes regime expressions. Beyond the Stokes regime, van Wachem et al. [28] and Ouchene et al. [29] studied a turbulent channel flow laden with ellipsoidal particles using LES and DNS, respectively, employing the flow coefficients developed by themselves in previous works.

Rosendahl group developed a model for the numerical computation of cylindrical and superellipsoidal particles in laminar and turbulent flows in the intermediate Reynolds numbers regime [30–32]. Particle angular velocity and orientation were computed by means of the Euler parameters. Using a linear relationship between the drag coefficient and the ellipsoid parameters, it was possible to establish a correlation valid up to Reynolds numbers of 1000. To estimate the influence of the orientation, a correlation between the maximum (90°) and minimum (0°) drag was employed. Drag force was calculated using the projected area perpendicularly to the flow. The lift force was expressed in function of the drag coefficient and particle orientation. Other lift forces, such as those due to the fluid velocity gradients or particle rotation, were not considered. The study case was a combustion chamber with straw particles, which were quite well approximated by cylinders. It was found that straw particles were better dispersed than spheres [30], a fact that properly illustrates the importance of the correct modeling of nonspherical particles motion. In a later work [32], other forces such as added mass and pressure force were also included. Drag coefficient was computed using the Ganser [15] correlation, making it possible to numerically compute the biomass combustion chamber.

This contribution aims to study the motion of nonspherical particles immersed in homogeneous isotropic turbulent (HIT) velocity fields built from kinematic simulation at moderate Reynolds numbers. Computations were performed in a tailored in-house code. Properties analyzed include the Lagrangian time autocorrelation function, the translational and rotational particle response, and preferential orientation of the nonspherical particles in the turbulent flow, all of them in terms of particle aspect ratio and inertia.

2. Governing equations

2.1 Coordinate systems

To build the trajectory of a regular nonspherical particle, it is necessary to solve for its translational as well as rotational motion. However, whereas translation is solved in an inertial frame, rotation is solved referred to the so-called particle frame. Thus, the relevant coordinate frames and the transformations between them have to be introduced.

Figure 1 illustrates, in the case of a cylindrical particle, the employed coordinate systems: $\mathbf{x} = [x\, y\, z]$ is the inertial frame; $\mathbf{x}' = [x'\, y'\, z']$ is the particle frame, whose origin is in the particle center of mass and its axes are the particle principal axes; and $\mathbf{x}'' = [x''\, y''\, z'']$ is the comoving frame, which has its origin at the same point than particle frame but its axes are parallel to the inertial frame axes. In the particle

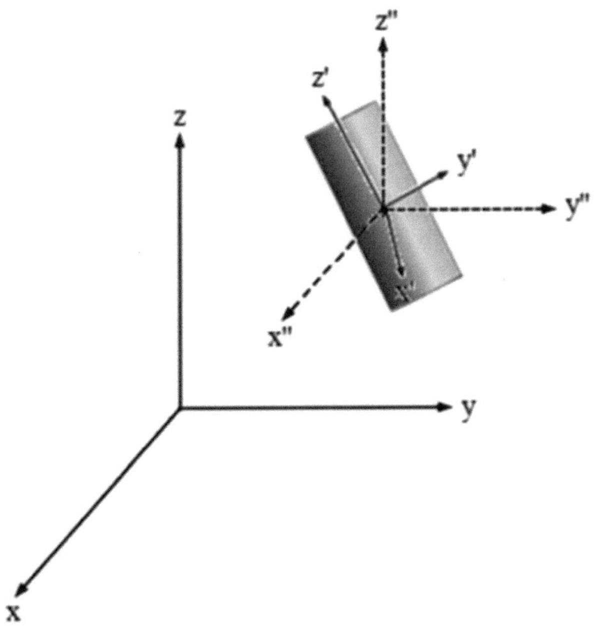

Figure 1.
Illustration of a cylindrical particle and the employed coordinate systems.

frame, the z' axis coincides with the particle symmetry axis and its position with respect to the comoving frame determines particle orientation.

Goldstein [33] gives the transformation between the comoving and particle coordinate systems, which is frequently employed in regular nonspherical particle tracking [31].

$$\mathbf{x}' = \mathbf{A} \cdot \mathbf{x}'' \tag{1}$$

A is the orthogonal matrix that performs the transformation. Its components are the direction cosines of the particle axes in the comoving frame, written in function of Euler angles (θ, ϕ, ψ). Such Euler angles are defined according to the x-convention of [33]:

$$\mathbf{A} = \begin{bmatrix} \cos\psi\cos\phi - \cos\theta\sin\phi\sin\psi & \cos\psi\sin\phi - \cos\theta\cos\phi\sin\psi & \sin\psi\sin\theta \\ -\sin\psi\cos\phi - \cos\theta\sin\phi\cos\psi & -\sin\psi\sin\phi + \cos\theta\cos\phi\cos\psi & \cos\psi\sin\theta \\ \sin\theta\sin\phi & -\sin\theta\cos\phi & \cos\theta \end{bmatrix} \tag{2}$$

The time evolution of such Euler angles depends on the particle angular velocity regarding the particle frame axes. However, there is a difficulty in the sense that such time evolution equations present an unavoidable singularity. Therefore, instead of the Euler angles, the Euler parameters $(\varepsilon_1, \varepsilon_2, \varepsilon_3, \eta)$ are used instead:

$$\varepsilon_1 = \cos\frac{\phi - \psi}{2}\sin\frac{\theta}{2}; \varepsilon_2 = \sin\frac{\phi - \psi}{2}\sin\frac{\theta}{2}; \varepsilon_3 = \sin\frac{\phi + \psi}{2}\cos\frac{\theta}{2}; \eta = \cos\frac{\phi + \psi}{2}\cos\frac{\theta}{2} \tag{3}$$

And the transformation matrix **A** is written as [33]:

$$\mathbf{A} = \begin{bmatrix} 1 - 2(\varepsilon_2^2 + \varepsilon_3^2) & 2(\varepsilon_1\varepsilon_2 + \varepsilon_3\eta) & 2(\varepsilon_1\varepsilon_3 - \varepsilon_2\eta) \\ 2(\varepsilon_1\varepsilon_2 - \varepsilon_3\eta) & 1 - 2(\varepsilon_1^2 + \varepsilon_3^2) & 2(\varepsilon_3\varepsilon_2 + \varepsilon_1\eta) \\ 2(\varepsilon_1\varepsilon_3 + \varepsilon_2\eta) & 2(\varepsilon_3\varepsilon_2 - \varepsilon_1\eta) & 1 - 2(\varepsilon_1^2 + \varepsilon_2^2) \end{bmatrix} \tag{4}$$

In the present study, the initial particle orientations are assigned by means of the Euler angles. From them, the corresponding Euler parameters are computed by Eq. (3), and with them, the initial transformation matrix is evaluated using Eq. (4). The Euler parameters evolve in time following Eq. (5), where the particle angular velocities are expressed in the particle frame of reference $\mathbf{x}' = [x'\ y'\ z']$.

$$\begin{bmatrix} \dfrac{d\varepsilon_1}{dt} \\ \dfrac{d\varepsilon_2}{dt} \\ \dfrac{d\varepsilon_3}{dt} \\ \dfrac{d\eta}{dt} \end{bmatrix} = \frac{1}{2}\begin{bmatrix} \eta\omega_{x'} - \varepsilon_3\omega_{y'} + \varepsilon_2\omega_{z'} \\ \varepsilon_3\omega_{x'} + \eta\omega_{y'} - \varepsilon_1\omega_{z'} \\ -\varepsilon_2\omega_{x'} + \varepsilon_1\omega_{y'} + \eta\omega_{z'} \\ -\varepsilon_1\omega_{x'} - \varepsilon_2\omega_{y'} - \varepsilon_3\omega_{z'} \end{bmatrix} \tag{5}$$

2.2 Particle motion equations

The nonspherical particle motion equations in a general fluid flow [34] are written as:

Translational motion:

$$m_p \frac{d\mathbf{u}_p}{dt} = \mathbf{F} \tag{6}$$

Rotational motion:

$$\begin{aligned} I_{x'} \frac{d\omega_{x'}}{dt} - \omega_{y'}\omega_{z'}\left(I_{y'} - I_{z'}\right) &= T_{x'} \\ I_{y'} \frac{d\omega_{y'}}{dt} - \omega_{x'}\omega_{z'}\left(I_{z'} - I_{x'}\right) &= T_{y'} \\ I_{z'} \frac{d\omega_{z'}}{dt} - \omega_{y'}\omega_{x'}\left(I_{x'} - I_{y'}\right) &= T_{z'} \end{aligned} \tag{7}$$

Here, m_p is the mass of the particle, $\mathbf{u}_p = \left[u_{px}\; u_{py}\; u_{pz}\right]$ is the translational velocity of the particle center of mass, referred to the inertial frame, and $\mathbf{F} = \left[F_x\; F_y\; F_z\right]$ is the external forces acting on the particle. The moments of inertia with respect to the particle frame axes are $\left[I_{x'}\; I_{y'}\; I_{z'}\right]$, and $\left[T_{x'}\; T_{y'}\; T_{z'}\right]$ are the torques experienced by the particle. It should be remarked that the equations for the translation motion are computed in the inertial frame but those of the rotation motion are expressed in the particle frame. In case of the torque experienced by the particle, it has two contributions: the pitching torque, due to the noncoincidence of the particle center of mass and center of pressure (same fact that happens in an airfoil), and the rotational torque, due to the viscous resistance experienced by a rotating body inside a fluid, generated by the differences between fluid and particle rotational velocities.

In addition to a sphere, the four ellipsoids of Zawstawny et al. [20] have been chosen. They have different sphericities and aspect ratio (see **Table 1**). In **Table 1**, *a* denotes the major semiaxis and *b* the minor semiaxis.

Using DNS for ellipsoidal particles immersed in a uniform flow, Zastawny et al. [20] determined correlations for the flow coefficients (drag C_D, lift C_L, pitching torque C_T, and rotational torque C_R). Such coefficients are written as [20]:

$$C_D = \frac{F_D}{\frac{1}{2}\rho\tilde{u}^2\frac{\pi}{4}d_p^2};\; C_L = \frac{F_L}{\frac{1}{2}\rho\tilde{u}^2\frac{\pi}{4}d_p^2};\; C_T = \frac{F_T}{\frac{1}{2}\rho\tilde{u}^2\frac{\pi}{8}d_p^3};\; C_R = \frac{F_R}{\frac{1}{2}\rho\left(\frac{d_p}{2}\right)^5|\boldsymbol{\Omega}|^2} \tag{8}$$

Here, d_p is the volume equivalent particle diameter or the diameter of a sphere with the same volume as the considered particle. The relative fluid velocity with respect to the particle is $\tilde{\mathbf{u}} = \mathbf{u} - \mathbf{u}_p$ and $\boldsymbol{\Omega} = \frac{1}{2}\ \times \mathbf{u} - \boldsymbol{\omega}_p$ is the fluid relative rotation with $\boldsymbol{\omega}_p$ being the particle angular velocity.

The developed correlations depend not only on particle Reynolds number $Re = \rho d_p\tilde{u}/\mu$ and particle rotation number $Re_R = \rho d_p^2|\boldsymbol{\Omega}|/\mu$, but also on orientation φ. They are written as [20]:

Shape	Aspect ratio	Sphericity
Ellipsoid 1 (prolate)	$\frac{a}{b} = \frac{5}{2}$	0.88
Ellipsoid 2 (prolate)	$\frac{a}{b} = \frac{5}{4}$	0.99
Disc (oblate)	$\frac{a}{b} = \frac{5}{1}$	0.62
Fiber (prolate)	$\frac{a}{b} = \frac{5}{1}$	0.73

Table 1.
Ellipsoids evaluated by Zastawny et al. [20]. a *and* b *are the major and minor semiaxis, respectively.*

$$C_D = C_{D,0} + (C_{D,90} - C_{D,0}) \sin^{a_0}\varphi; C_{D,0} = \frac{a_1}{Re^{a_2}} + \frac{a_3}{Re^{a_4}}; C_{D,90} = \frac{a_5}{Re^{a_6}} + \frac{a_7}{Re^{a_8}} \quad (9)$$

a values are listed in [20] as also coefficients b, c, r:

$$C_L = \left(\frac{b_1}{Re^{b_2}} + \frac{b_3}{Re^{b_4}}\right)(\sin\varphi)^{b_5+b_6Re^{b_7}}(\cos\varphi)^{b_8+b_9Re^{b_{10}}} \quad (10)$$

$$C_T = \left(\frac{c_1}{Re^{c_2}} + \frac{c_3}{Re^{c_4}}\right)(\sin\varphi)^{c_5+c_6Re^{c_7}}(\cos\varphi)^{c_8+c_9Re^{c_{10}}} \quad (11)$$

$$C_R = r_1 Re_R^{r_2} + \frac{r_3}{Re_R^{r_4}} \quad (12)$$

Moreover, for the cylinders, Vakil and Green [19] developed correlations for drag and lift coefficients depending on orientation, Reynolds number based on its diameter Re_D, and aspect ratio AR. In this case, such coefficients are expressed in terms of cylinder length L and diameter D:

$$C_{D,cyl} = \frac{F_D}{\frac{1}{2}\rho\tilde{u}^2 LD}; C_{L,cyl} = \frac{F_L}{\frac{1}{2}\rho\tilde{u}2LD} \quad (13)$$

The correlations are expressed as:

$$C_{L,cyl}(\varphi, AR, Re_D) = A_2(AR, Re_D)\sin 2\varphi + A_4(AR, Re_D)\sin 4\varphi \quad (14)$$

$$\frac{C_{D,cyl}}{C_{D\perp}}(\varphi, AR, Re_D) = A_1(AR, Re_D)\cos 2\varphi + A_0(AR, Re_D); C_{D\perp}(AR, Re_D) = \left(\kappa_1 + \frac{\kappa_2}{AR}\right)Re_D^{\left(\kappa_3 + \frac{\kappa_4}{AR}\right)} \quad (15)$$

Coefficients κ and those γ in functions A_i, $i = 0, 1, 2, 4$:

$$A_i(AR, Re_D) = \beta_{i1}(AR)\ln Re_D + \beta_{i2}(AR); \beta_{ij}(AR) = \gamma_{ij1} + \gamma_{ij2}\exp\left(\gamma_{ij3}AR\right) \quad j = 1, 2 \quad (16)$$

can be found in Vakil and Green [19].

However, expressions for the pitching and rotational torque coefficients are not provided in [19]. Therefore, for the cylinders, the approach of [31] has been assumed. In [31], the distance between the center of mass and the center or pressure in a cylinder, l_{CP}, in terms of AR and φ, was proposed to be:

$$l_{CP} = 0.25\frac{L}{2}\left(1 - e^{3(1-AR)}\right)|\cos^3\varphi| \quad (17)$$

Then, the pitching torque $\mathbf{T}_P$ is just the cross-product between the particle orientation unitary vector and the resultant force acting on it times l_{CP}. Nevertheless, this torque is computed in the inertial frame of reference, so it should be transformed to the particle frame before being included in Eq. (7) to calculate the particle angular velocity. The approach to compute the viscous rotational torque $\mathbf{T}_R$ is to integrate along the particle length the torque due to the drag force with respect the particle center of mass and it is described in [31]. This torque is given directly in the particle frame.

Particle motion equations and correlations for cylinders and ellipsoids presented in this section have been implemented in an in-house code. The numerical integration of the ordinary differential equations that govern the motion of nonspherical particles has been performed by a fourth-order Runge-Kutta method, with small enough time steps to avoid numerical instabilities [35, 36]. The fluid velocity field in which particles are immersed has been built by the kinematic simulation technique described in the next section. It is known that Runge-Kutta methods do not satisfy the time-reversal property, a fact that makes such methods inappropriate for integrating energy-conserving systems, for instance. However, particle equations are dissipative systems (as they include viscous drag forces) and, for them, Runge-Kutta algorithms can be used [37] provided that the time step is small enough to keep the errors bounded.

3. Kinematic simulation

There exist different options to calculate the Lagrangian properties in a turbulent flow. The starting point is the trajectory equation in which the position $\mathbf{x}(\mathbf{x}_0, t)$ of a particle released at point $\mathbf{x}_0$ at time $t = 0$ is calculated solving:

$$\frac{d\mathbf{x}}{dt} = \mathbf{u}(\mathbf{x}, t) \tag{18}$$

Here, $\mathbf{u}(\mathbf{x}, t)$ is the Eulerian velocity field. If it is known, it is possible to solve Eq. (18); however, finding $\mathbf{u}(\mathbf{x}, t)$ is not an easy task. One possibility is to work with Lagrangian statistics but then it would be needed to close the relevant Lagrangian correlations. Another option to solve Eq. (18) is to use DNS to obtain $\mathbf{u}(\mathbf{x}, t)$; however, this is computationally very expensive. A much more economical alternative is the use of kinematic simulation (KS) to compute the Lagrangian characteristics of turbulent flow fields. In this technique, stochastic fluid velocity fields are constructed in such a way that their statistical properties are in agreement with those extracted from experiments or reliable DNS. The main advantage of KS is that it employs an explicit continuous formula for computing $\mathbf{u}(\mathbf{x}, t)$, so it is not needed to perform interpolation of the fluid velocity field. Moreover, KS results of two particle statistics in HIT have been validated versus DNS showing good agreement [38].

The three-dimensional Eulerian velocity field to be employed in Eq. (18) is built as a series of random Fourier modes. The velocity field is solenoidal at each realization by construction. Moreover, the energy spectrum of the Fourier modes is prescribed, for example, by a power law, so the effects of small flow scales on Lagrangian statistics are directly included. Such KS velocity field is written as [39]:

$$\mathbf{u}(\mathbf{x}, t) = \sum_{n=1}^{N} \mathbf{A}_n \cos\left(\mathbf{k}_n \bullet \mathbf{x} + \omega_n t\right) + \mathbf{B}_n \sin\left(\mathbf{k}_n \bullet \mathbf{x} + \omega_n t\right) \tag{19}$$

$\mathbf{k}_n$ represents the n-th wave number; coefficients $\mathbf{A}_n$, $\mathbf{B}_n$ are random, uncorrelated vectors perpendicular to $\mathbf{k}_n$, whose amplitudes are chosen according to the prescribed energy spectrum $E(k)$ [39]. Here, the energy spectrum has been the Kolmogorov decay law of $-5/3$.

ω_n is the n-th frequency, which determines the unsteadiness of the corresponding mode; it is written proportional to the eddy-turnover time of the n-th mode:

$$\omega_n = \lambda\sqrt{k_n^3 E(k_n)} \tag{20}$$

Here, λ is a parameter of order 1 that governs the unsteadiness of the velocity field. In three-dimensional HIT flows, it has been demonstrated [38] that the

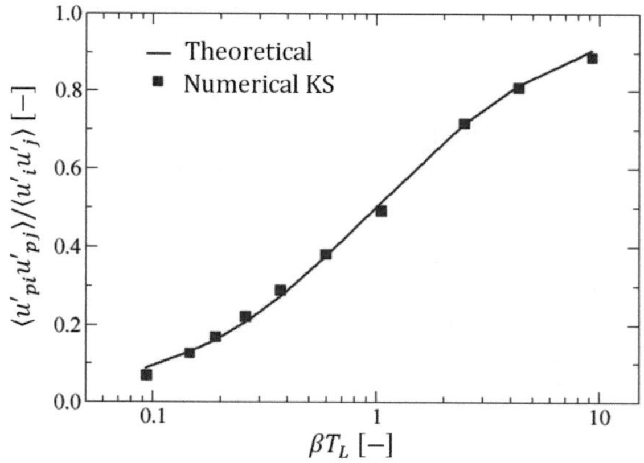

Figure 2.
Comparison of spherical particle Reynolds stresses, obtained with KS versus theoretical values for i = j.

statistical characteristics of two-particle diffusion are independent of λ. In particular, in this work, two values of the unsteadiness parameter of 0 and 0.5 have been tested, without significant differences in the computed statistical properties. Therefore, following the suggestions of [39], the value 0.5 has been adopted in the present computations.

Because of the construction of the velocity field given by Eq. (19), $\mathbf{k}_n \bullet \mathbf{A}_n = \mathbf{k}_n \bullet \mathbf{B}_n = 0$, it is solenoidal trajectory by trajectory. Moreover, as shown in [40], such field includes in each realization turbulent-like patterns as eddying, straining, and streaming regions.

To validate the spherical particles tracking in the KS velocity field, the values of particle Reynolds stresses (RS) in HIT have been selected. In this configuration, Hyland et al. [41] demonstrated that, as the fluid turbulence is homogeneous, particle RS only depend on time and they can be written as $\langle u'_{pi}u'_{pj}(t)\rangle/\langle u'_iu'_j\rangle = q(t)\delta_{ij}$, that is, they are an isotropic tensor. Moreover, in the asymptotic limit, $q(t \rightarrow \infty) = \beta T_L/(1 + \beta T_L)$, where β is the inverse of particle relaxation time (see Eq. (21) below) and T_L is the Lagrangian time scale of fluid turbulence. **Figure 2** presents the numerical results for particle RS computed with KS and the asymptotic expression $q(t \rightarrow \infty)$. As **Figure 2** readily shows, the asymptotic particle RS are very well reproduced by the numerical particle tracking in the KS velocity field in the range of two decades for βT_L.

4. Numerical simulation

Computations were performed in a tailored in-house code. The turbulent velocity field generated with KS resembles one of the fields worked in [39]. Such velocity field is characterized by a fluctuating velocity *u'* = 1 m/s, a fluid Reynolds number of 10^4 resulting in a Kolmogorov length scale $\eta_K \approx 6.286$ mm, associated Kolmogorov time scale $\tau_K \approx 10$ ms, and a fluid integral Lagrangian time scale of turbulence $T_L = 0.56\,s$. Those values are matched by the present KS.

The regular nonspherical particles studied have been the ellipsoids in [20] and the cylinders in [19]. In all cases, particles have the same particle volume equivalent diameter $d_p = 200\ \mu m$, hence much smaller than η_K. Therefore, such particles can be thought as immersed in a uniform flow field. The Stokes number has been modified by adjusting the material density of particles being the Stokesian particle relaxation time defined as:

$$\tau_p = \beta^{-1} = \frac{\rho_p d_P^2}{18\mu} \tag{21}$$

If the Kolmogorov time scale τ_K is taken as the fluid time scale, particle Stokes number is defined as $St = \tau_p/\tau_K$. According to this nondimensional number, three particle inertia classes are considered: light ($St \approx 0.5$), intermediate ($St \approx 10$), and heavy ($St \approx 100$). However, as cases with $Re > 1$ are considered, an effective particle relaxation time is introduced as $\tau_{p,eff} = \tau_p/ReC_D$, allowing the introduction of an effective Stokes number $St_{eff} = \tau_{p,eff}/\tau_K$. Therefore, the values of such effective Stokes number are $St_{eff} \approx$ 0.3 (light), 5 (intermediate), and 40 (heavy).

Simulations proceed in the following way: for each KS realization of HIT fluid velocity field, a particle is located in the center of the domain with zero initial velocity; particle translational and rotational motion is computed from Eqs. (6) and (7), its orientation is calculated from Eq. (1), and its trajectory is built; particle tracking lasts for around 10 fluid integral time scales; and particle properties are stored every second for evaluation. Such process is carried out a sufficient number of times to reach significant statistical results. In this study, statistics has been performed based on 10^5 KS realizations.

In the following section, the results of the particle Lagrangian time autocorrelation function, the translational and rotational particle response, and preferential orientation of the nonspherical particles in the turbulent flow are analyzed as function of their shape and effective Stokes number.

5. Results and discussion

The Lagrangian autocorrelation function $R_{L,t}(\tau)$ for translational motion is expressed as:

$$R_{L,t}(\tau) = \frac{\langle \mathbf{u}_p(0) \bullet \mathbf{u}_p(\tau) \rangle}{\langle \mathbf{u}_p \bullet \mathbf{u}_p(0) \rangle} \tag{22}$$

τ represents the time delay. With the objective of making results independent of particle injection conditions, statistics are started to be collected after 2 s. The obtained results for the Lagrangian autocorrelation function (LAF) of the ellipsoids of Zastawny et al. [20] are presented in the left part of **Figure 3**, including the results for spherical particles, whereas those of the cylinders of Vakil and Green [19] are in the right side of such figure. Horizontal axis is the nondimensional time delay, τ/T_L. As the lighter particles have an autocorrelation function nearly equal to that of the fluid (tracer limit), the corresponding curves are not shown in **Figure 3**. Therefore, only the curves for intermediate and high inertia particles are presented. Moreover, in **Figure 3** also the fluid Lagrangian (brown curve) and Eulerian (cyan curve) $R_{L,t}(\tau)$'s are included for comparison.

As it can be readily seen from **Figure 3**, higher inertia particles are characterized by larger integral Lagrangian time scales (ILTSs) (defined as the integral up to infinity of $R_{L,t}(\tau)$), as a result of their smaller responsiveness to the turbulent fluctuations. Same as in [23], all particle LAFs are mainly in between the fluid LAF and Eulerian autocorrelation function (EAF). As a consequence of inertia, for the smallest values of τ, the heavy $R_{L,t}(\tau)$ overcomes the fluid EAF, differently from [23] who considered only noninertial particles.

Moreover, for intermediate inertia, the curves for all particle shapes nearly collapse in a single curve. On the other hand, a shape effect is noticeable for the

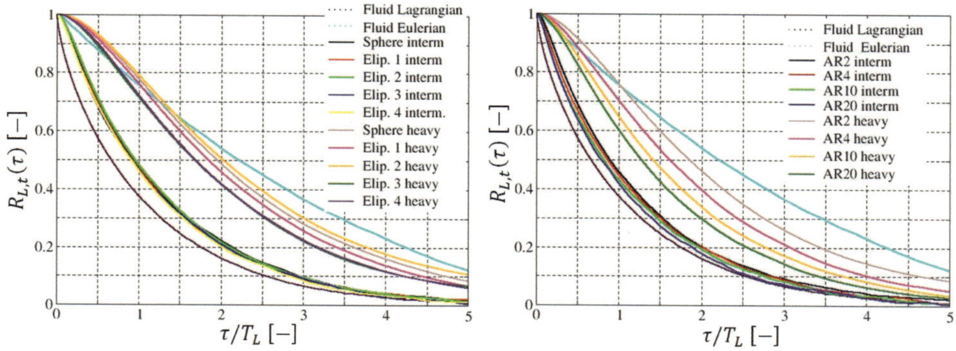

Figure 3.
Computed $R_{L,t}(\tau)$ curves for ellipsoidal particles [20] (left) and cylindrical particles [19] (right). Fluid Lagrangian and Eulerian curves are included for comparison.

heaviest particles, where the various shapes present differences in their curves. It is interesting to realize that ILTSs of higher aspect ratio (AR) are below those of smaller AR, for both ellipsoids and cylinders. This effect is a Reynolds number effect due to the dependence of drag coefficient on shape and AR: an interaction between translation and rotation motions occurs that results in a spreading of the particle effective Stokes number. As a consequence, particles with higher Reynolds numbers also have larger effective inertia (reflected on an increased Stokes number) and, therefore, their LAF decreases slower, implying a higher ILTS. In the ellipsoids case, it happens that those of typ. 2 present a $R_{L,t}(\tau)$ curve slightly over that of the spherical particle, as they have lower effective Stokes number. Also, the LAF curve for the disc-like particles in this case is very similar to that of the fiber.

In an analogous way to translational LAF, a rotational autocorrelation function (RAF) $R_{L,r}(\tau)$ can be defined in terms of the time delay τ as:

$$R_{L,r}(\tau) = \frac{\langle \boldsymbol{\omega}_p(0) \bullet \boldsymbol{\omega}_p(\tau) \rangle}{\langle \boldsymbol{\omega}_p \bullet \boldsymbol{\omega}_p(0) \rangle} \tag{23}$$

where the particle angular velocity is denoted by $\boldsymbol{\omega}_p$.

The obtained results for the Lagrangian rotational autocorrelation function of the ellipsoids of Zastawny et al. [20] are shown in the left part of **Figure 4**, whereas those of the cylinders of Vakil and Green [19] are in the right side of such figure. Again, horizontal axis is the nondimensional time delay, τ/T_L. Similar to the case of translational motion, the angular velocities of heavy particles keep correlated for longer times than those of lighter particles. Such correlation time for ellipsoidal particles is much shorter than that of the translational motion. Also, for all inertia cases, the RAF of disc-like particles drops quicker than for the prolate ellipsoids. As mentioned for the translational correlations, the RAF curves for the prolate ellipsoids collapse for the lighter particles, but they show noticeable differences for the intermediate and large inertia particles demonstrating an effect of the aspect ratio on $R_{L,r}(\tau)$. Ellipsoid 2, with the smaller AR, has the higher RAF curve of all prolate ellipsoids, while Ellipsoid 1 and the fiber have very similar rotational correlation functions.

In the case of cylinders (**Figure 4**, right), the RAF curves for all AR and inertias are different. For the smallest inertia particles, RAF decreases with increasing AR, similar to what was found for LAFs in the translational motion. Moreover, the $R_{L,r}(\tau)$ curve presents negative values for the two largest aspect ratios of 10 and 20. For the intermediate particles, the RAF curves keep the same decreasing trend with increasing aspect ratio as the light particles; however, in this case, correlation times

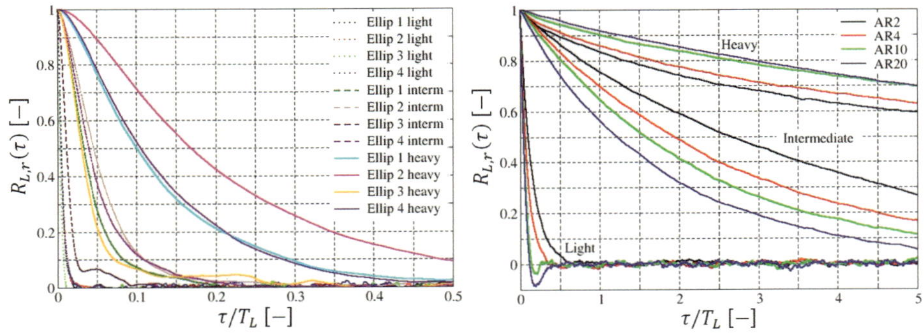

Figure 4.
Computed $R_{L,r}(\tau)$ curves for ellipsoidal particles [20] (left) and cylindrical particles [19] (right).

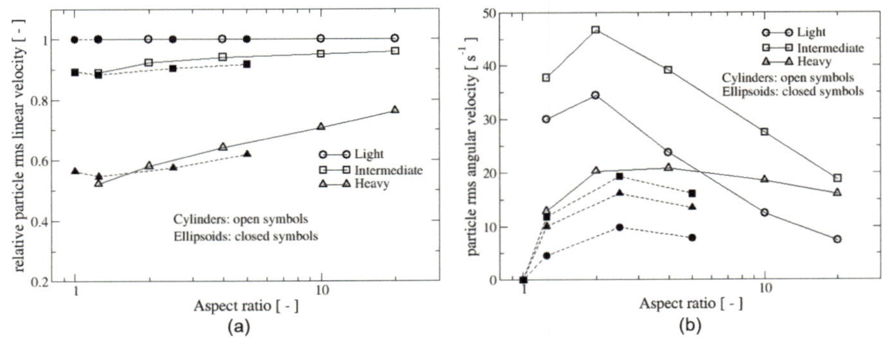

Figure 5.
Relative particle rms of particle linear velocity (a) and angular rms velocity (b). In all cases, the dependence on aspect ratio and inertia is considered. Closed symbols refer to prolate ellipsoids and open symbols to cylinders.

for angular velocities are very much increased and they are significantly higher than for ellipsoids. The previous trend is reversed for the heavy particles as the RAF curves augment with increasing *AR*; however, as **Figure 4** (right) suggests, an asymptotic value for *L*/*D* seems to exist because the curves for *AR* = 10 and 20 are very close to each other. The change of behavior of the RAF with increasing inertia could be due to the fact that for the small and intermediate inertia, the particle relaxation time for rotation reduces with growing *AR*, whereas for heavy particles, such relaxation time behaves in the opposite way. Nevertheless, this fact must be further investigated, possibly using fully resolved simulations.

Next, the response of the nonspherical particles to the fluid fluctuating velocities is analyzed for both translation and rotation motions. **Figure 5(a)** shows the behavior of the particle's relative linear root mean square (rms) velocity, that is, u_p'/u', where u' is the fluid rms fluctuating translational velocity and u_p' denotes the same quantity but for the particles. The aspect ratio is in the horizontal axis, whereas the different curves correspond to the various inertia cases. On the one side, as it could be anticipated, u_p' reduces with increasing particle inertia because the more inertial particles are not able to follow all fluid velocity fluctuations. In fact, as it was found for the LAF, the less inertial particles present, for both ellipsoids and cylinders and for all values of *AR*, the same fluctuating velocities as the fluid, indicating that they behave as fluid tracers.

As inertia increases, u_p'/u' decreases monotonically, as expected. However, it increases with growing aspect ratio for both cylinders and ellipsoids. There is one exception that spherical particles have values of u_p'/u' slightly above those of the ellipsoid with AR = 1.25. The trend of increasing particle fluctuating velocities with aspect ratio is consistent with the aforementioned fact that effective Stokes number tends to reduce with growing AR; therefore, particles with lower AR respond less to the fluid fluctuations than the more elongated ones. This result has been obtained for particles much smaller than Kolmogorov length scale; therefore, such particles can be considered to be immersed in a uniform flow field, just in the same conditions as the flow coefficient correlations were developed. In the study of Hölzer and Sommerfeld [42], nevertheless, a different result was found. Using a DNS based on the lattice Boltzmann method (LBM), Hölzer and Sommerfeld [42] obtained that relative particle fluctuating velocity reduced with increasing AR. The main difference with the present work is that Hölzer and Sommerfeld [42] employed particles with size well above η_K. The authors explained the fact arguing that particles averaged the fluid fluctuations on their surface, which augmented with increasing AR. Let us remark that both results are not conflicting as the size range of the employed particles in the two studies is very different. Further work combining DNS with nonspherical point particles smaller than η_K is necessary to explain this point.

Figure 5(b) presents the behavior of the particle angular rms velocity, ω'_p. In homogeneous and isotropic turbulence, the angular velocity of spherical particles is zero because of viscous damping and the absence of pitching torque. The situation is different in nonspherical particles because in them the geometrical and pressure centers do not coincide, so a net pitching torque is produced that promotes nonzero angular velocities. As illustrated in **Figure 5(b)**, ω'_p increases with AR and reaches up to a maximum of $AR \approx 2$ and decreases with higher values of aspect ratio. The shape of the curve is the same for ellipsoids and cylinders. Such behavior was also found in [42], and it was explained observing that, with increasing AR, the moment of inertia along the major axis reduces and along the minor axis increases, which would lead to higher and lower ω'_p, respectively. On the other hand, and similar to what happened with u'_p, for inertial particles, ω'_p decreases as inertia augments.

In the following, the correlation relative velocity direction-particle orientation is analyzed depending on inertia and aspect ratio. A well-known fact is that regular nonspherical particles falling through a still liquid at intermediate Reynolds numbers tend to be oriented in a determined direction. Cylinders and prolate ellipsoids are prone to keep their symmetry axis (z' in **Figure 1**) perpendicular to the flow, thus maximizing drag. Differently, discs and oblate ellipsoids tend to move with the symmetry axis aligned with the flow, also maximizing drag [43]. However, spheroidal Stokes particles only show a preferential orientation if a persistent velocity gradient exists [27]. Therefore, in HIT flow where there are no mean velocity gradients, a Stokes particle will not have any preferred orientation.

Newsom and Bruce [44] analyzed the influence of turbulence on preferential alignment of quite elongated fibers with $Re \approx 1$. As explained by [44], preferential orientation of such fibers falling through a still fluid can only be clarified if fluid inertial effects are considered. In the Stokes regime, Khayat and Cox [43] demonstrate that the force distribution on the fiber is symmetrically distributed along its axis, independent of its orientation regarding the flow and, as a result, the fiber experiences a zero net torque. Beyond the Stokes regime, $Re > 1$, when the fiber has an oblique orientation with respect the flow, such distribution of the force is not any more symmetric and it experiences a net pitching torque. Such torque will promote a rotation that drives the fiber to adopt an orientation where its symmetry axis is orthogonal to the relative flow. Interestingly, if the fiber is oriented orthogonal or parallel to the flow, the net experienced torque is zero, due to the symmetry of the force distribution; however, in the first case, this situation is stable, while in the second case, where the centers of gravity and pressure do not coincide, this situation is unstable.

Previous reasoning is valid too for another kind of nonspherical shapes as disc-like, cylindrical, or ellipsoidal [45]. For high Reynolds numbers, that is, $Re > 100$ appears a secondary motion overimposed to the particles predominant movement direction. Such secondary motion is promoted by a wake instability and vortex detachment from the rear surface of the particles. Two main kinds of secondary motion can be observed: large quasi-periodic swings along the main path, and a more or less chaotic tumbling forming a definite angle with the main motion direction. There is a coupling between this kind of oscillatory motion and the wake instability [10]: a vortex detachment follows at the end of a particle swing. Nevertheless, in the present study, such secondary motions do not appear as the considered particle Reynolds number is not large enough, that is, $Re < 40$.

Let now θ be the angle formed by the relative velocity, $\mathbf{u} - \mathbf{u}_p$, and the particle symmetry axis, z'. Therefore, $\cos\theta$ can be used to determine the orientation of the nonspherical particle with respect to the relative flow. In this work, particle preferential orientation is determined computing $|\cos\theta|$ along the trajectories of 10^5 particles. Computed values of $|\cos\theta|$ are sorted in equally distributed bins between 0 (particle axis orthogonal to relative velocity) and 1 (alignment between particle axis and relative velocity), and the corresponding probability density functions (Pdfs) are determined. Such Pdfs are shown in **Figure 6** in terms of $|\cos\theta|$. **Figure 6(a)** shows the results for the prolate ellipsoids in terms of particle inertia,

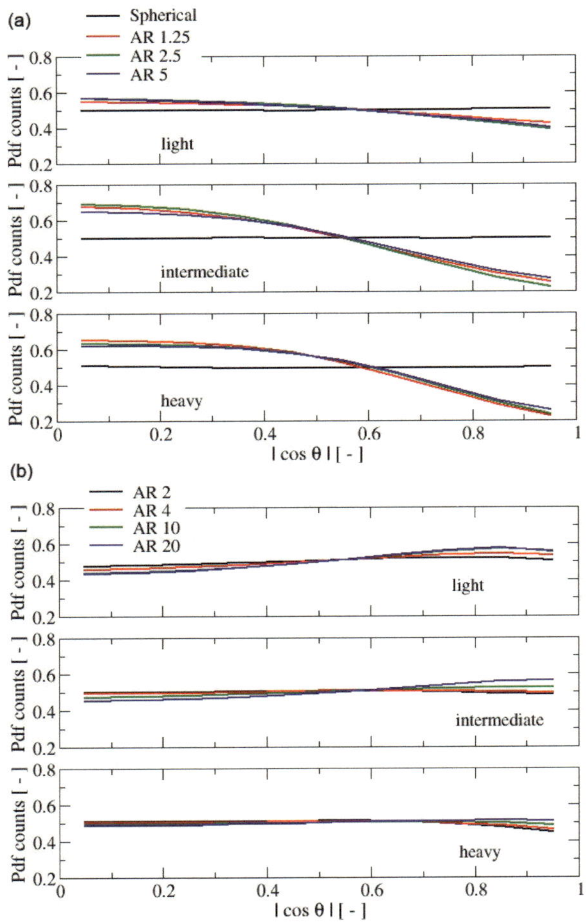

Figure 6.
Orientations probability density functions (Pdfs) of prolate ellipsoids (a) and cylinders (b) regarding the relative flow direction.

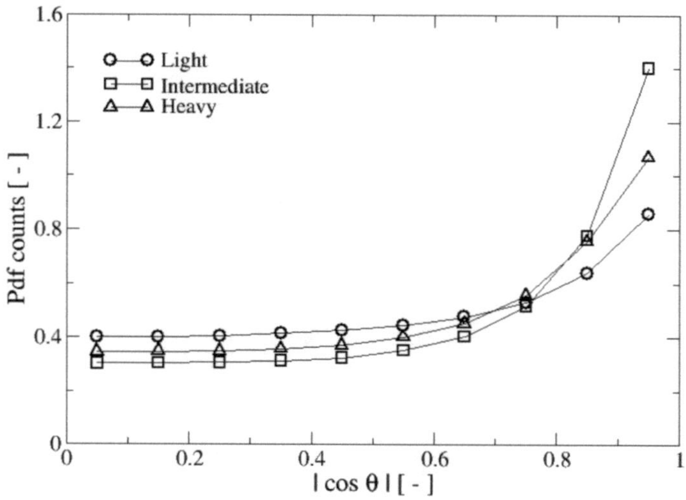

Figure 7.
Orientation probability density functions (Pdfs) of disc-like particles regarding the relative flow direction.

and **Figure 6(b)** presents the curves for the cylinders also depending on their inertia. Each inertia class is plotted in a separated frame. For the discs case, results are presented in **Figure 7**.

As it is observed in **Figure 6(a)**, it is found that prolate ellipsoids do manifest preferential orientation with respect to the relative velocity. Of course, spherical particles do not have a preferred orientation and the corresponding Pdf is a horizontal line (black color). Prolate ellipsoids have a preference for orientating its symmetry axis orthogonal to the relative flow, tending to maximize the drag, similar to what occurs in particle sedimentation studies. The orientation preference increases with inertia, which is quite similar for all aspect ratios considered in this study.

On the other hand, as it is presented in **Figure 6(b)**, cylinders seem not to have any preferred orientation in the HIT KS velocity field, being all the curves pretty flat. Only for the case of higher AR and lowest inertia, the curves show a trend to be slightly higher for values of $|\cos\theta|$ closer to one than to zero. Such result is qualitatively similar to the DNS computations of [27] in the central region of the channel.

For the discs, **Figure 7** shows that there is a clear trend of the particle symmetry axis to be aligned with the relative flow, again maximizing drag, similar to the results obtained for sedimenting particles in stagnant fluid. Such trend is more marked when particle inertia increases.

6. Conclusions

In this study, regular nonspherical particle responsiveness to HIT flows has been investigated in combination with KS of fluid velocity field. The main results obtained are the following: the particle LAF reduces when particle AR is augmented, because effective particle inertia decreases if aspect ratio increases; this is true for both translational and rotational time autocorrelation functions. In the case of cylinders, $R_{L,r}(\tau)$ is much higher than for ellipsoids, a fact that requires further clarification through particle resolved simulations. Additionally, the fluctuating particle velocity increases for growing AR in the considered case of particles much smaller than Kolmogorov length scale; such behavior is contrary, although not conflicting, to the findings of [42] for fully resolved particles with sizes larger than the Kolmogorov length scale. For both ellipsoids and cylinders, the particle angular rms velocity first increases with aspect ratio, reaches a maximum of $AR \approx 2$, and then decreases again, which is explained because with increasing aspect ratio, the moment of inertia around the longitudinal axis decreases and around the radial axis increases, which would lead to higher and lower rms angular velocities, respectively. Finally, in agreement with Marchioli et al. [27], cylinders seem not to prefer any specific orientation in the KS HIT velocity field; however, prolate ellipsoids tend to be oriented with its symmetry axis orthogonal to the relative flow, maximizing the drag. Oblate ellipsoids and disc-like particles also show a preferential orientation, tending to align their symmetry axis with the relative flow velocity.

Acknowledgements

The financial support of Universidad Autónoma de Occidente is gratefully acknowledged.

Author details

Santiago Laín
Department of Energetics and Mechanics, PAI+ Group, Universidad Autónoma de Occidente, Cali, Colombia

*Address all correspondence to: slain@uao.edu.co

References

[1] Simonin O. Statistical and continuum modelling of turbulent reactive particulate flows, Part II: Application of a two-phase second-moment transport model for prediction of turbulent gas-particle flows. In: Von Karman Institute for Fluid Mechanics Lecture Series, 2000-6. 2000

[2] Sommerfeld M, Lain S. From elementary processes to the numerical prediction of industrial particle-laden flows. Multiphase Science and Technology. 2009;**21**:123-140

[3] Sommerfeld M, Lain S. Stochastic modelling for capturing the behaviour of irregular-shaped non-spherical particles in confined turbulent flows. Powder Technology. 2018;**332**:253-264

[4] Jeffery G. The motion of ellipsoidal particles immersed in a viscous fluid. Proceedings of the Royal Society. 1922; **102A**:161-179

[5] Happel J, Brenner H. Low Reynolds Number Hydrodynamics. 2nd ed. The Hague: Martinus Nijhoff; 1983. 553 p

[6] Blaser S. Forces on the surface of small ellipsoidal particles immersed in a linear flow field. Chemical Engineering Science. 2002;**57**:515-526

[7] Squires L, Squires W Jr. The sedimentation of thin discs. Transaction of the American Institute of Chemical Engineers. 1937;**33**:1-12

[8] Pettyjohn ES, Christiansen EB. Effect of particle shape on free-settling rates of isometric particles. Chemical Engineering Progress. 1948;**44**:157-172

[9] Heiss JF, Coull J. The effect of orientation and shape on the settling velocity of non-isometric particles in a viscous medium. Chemical Engineering Progress. 1952;**48**:133-140

[10] Willmarth WW, Hawk NE, Harvey RL. Steady and unsteady motions and wakes of freely falling disks. Physics of Fluids. 1964;**7**:197-208

[11] McKay G, Murphy WR, Hills M. Settling characteristics of discs and cylinders. Chemical Engineering Research and Design. 1988;**66**:107-112

[12] Haider A, Levenspiel O. Drag coefficient and terminal velocity of spherical and nonspherical particles. Powder Technology. 1989;**58**:63-70

[13] Thompson TL, Clark NN. A holistic approach to particle drag prediction. Powder Technology. 1991;**67**:57-66

[14] Swamee PK, Ojha CAP. Drag coefficient and fall velocity of nonspherical particles. Journal of Hydraulic Engineering. 1991;**117**: 660-667

[15] Ganser GH. A rational approach to drag prediction of spherical and nonspherical particles. Powder Technology. 1993;**77**:143-152

[16] Tran-Cong S, Gay M, Michaelides EE. Drag coefficients of irregularly shaped particles. Powder Technology. 2004;**139**:21-32

[17] Hölzer A, Sommerfeld M. New and simple correlation formula for the drag coefficient of non-spherical particles. Powder Technology. 2008;**184**:371-365

[18] Hölzer A, Sommerfeld M. Lattice Boltzmann simulations to determine drag, lift and torque acting on non-spherical particles. Computers and Fluids. 2009;**38**:572-589

[19] Vakil A, Green SI. Drag and lift coefficients of inclined finite circular cylinders at moderate Reynolds

numbers. Computers and Fluids. 2009; **38**:1771-1781

[20] Zastawny M, Mallouppas G, Zhao F, van Wachem B. Derivation of drag and lift force and torque coefficients for non-spherical particles in flows. International Journal of Multiphase Flow. 2012;**39**:227-239

[21] Ouchene R, Khalij M, Arcen B, Tanière A. A new set of correlations of drag, lift and torque coefficients for non-spherical particles and large Reynolds numbers. Powder Technology. 2016;**303**:33-43

[22] Fan FG, Ahmadi G. Dispersion of ellipsoidal particles in an isotropic pseudo-turbulent flow field. Transactions of the ASME, Journal of Fluids Engineering. 1995;**117**:154-161

[23] Olson JA. The motion of fibres in turbulent flow, stochastic simulation of isotropic homogeneous turbulence. International Journal of Multiphase Flow. 2001;**27**:2083-2103

[24] Lin J, Shi X, Yu Z. The motion of fibers in an evolving mixing layer. International Journal of Multiphase Flow. 2003;**29**:1355-1372

[25] Zhang H, Ahmadi G, Fan FG, McLaughlin JB. Ellipsoidal particles transport and deposition in turbulent channel flows. International Journal of Multiphase Flow. 2001;**27**:971-1009

[26] Mortensen PH, Andersson HI, Gillissen JJJ, Boersma BJ. Dynamics of prolate ellipsoidal particles in a turbulent channel flow. Physics of Fluids. 2008;**20**:093302

[27] Marchioli C, Fantoni M, Soldati A. Orientation, distribution and deposition of elongated, inertial fibers in turbulent channel flow. Physics of Fluids. 2010;**22**: 033301

[28] van Wachem B, Zastawny M, Zhao F, Malloupas G. Modelling of gas–solid turbulent channel flow with non-spherical particles with large stokes numbers. International Journal of Multiphase Flow. 2015;**68**:80-92

[29] Arcen B, Ouchene R, Kahlij M, Tanière A. Prolate spheroidal particles' behavior in a vertical wall-bounded turbulent flow. Physics of Fluids. 2017; **29**:093301

[30] Rosendahl L. Using a multi-parameter particle shape description to predict the motion of non-spherical particle shapes in swirling flow. Applied Mathematical Modelling. 2000;**24**:11-25

[31] Yin C, Rosendahl L, Kaer SK, Sorensen H. Modelling the motion of cylindrical particles in a nonuniform flow. Chemical Engineering Science. 2003;**58**:3489-3498

[32] Yin C, Rosendahl L, Kaer SK, Condra TJ. Use of numerical modelling in design for co-firing biomass in wall-fired burners. Chemical Engineering Science. 2004;**59**:3281-3292

[33] Goldstein H. Classical Mechanics. 2nd ed. Vol. 793. New York: Addison-Wesley; 1980

[34] Gallily I, Cohen AH. On the orderly nature of the motion of nonspherical aerosol particles II. Inertial collision between a spherical large droplet and an axially symmetrical elongated particle. Journal of Colloid and Interface Science. 2979;**68**:338-356

[35] Göz MF, Lain S, Sommerfeld M. Study of the numerical instabilities in Lagrangian tracking of bubbles and particles in two-phase flow. Computers and Chemical Engineering. 2004;**28**: 2727-2733

[36] Göz MF, Sommerfeld M, Lain S. Instabilities in Lagrangian tracking of

bubbles and particles in two-phase flow. AICHE Journal. 2006;**52**:469-477

[37] Thijssen MJ. Computational Physics. 2nd ed. Cambridge: Cambridge University Press; 2007. 620 p

[38] Malik NA, Vassilicos JC. A Lagrangian model for turbulent dispersion with turbulent-like flow structure: Comparison with DNS for two-particle statistics. Physics of Fluids. 1999;**11**:1572-1580

[39] El-Maihy A. Study of diffusion and dispersion of particles using kinematic simulation [thesis]. Sheffield: University of Sheffield; 2003

[40] Davila J, Vassilicos JC. Richardson pair diffusion and the stagnation point structure of turbulence. Physical Review Letters. 2003;**91**:144501

[41] Hyland KE, McKee S, Reeks MW. Exact analytic solutions to turbulent particle flow equations. Physics of Fluids. 1999;**11**:1249-1261

[42] Hölzer A, Sommerfeld M. Analysis of the behaviour of cylinders in homogeneous isotropic turbulence by lattice Boltzmann method. ERCOFTAC Bulletin. 2010;**82**:11-16

[43] Khayat RE, Cox RG. Inertial effects on the motion of long slender bodies. Journal of Fluid Mechanics. 1989;**209**: 435-462

[44] Newsom RK, Bruce CW. Orientational properties of fibrous aerosols in atmospheric turbulence. Journal of Aerosol Science. 1998;**29**: 773-797

[45] Mandø M, Rosendahl L. On the motion of non-spherical particles at high Reynolds number. Powder Technology. 2010;**202**:1-13

Chapter 3

An Eulerian-Lagrangian Coupled Model for Droplets Dispersion from Nozzle Spray

Carlos G. Sedano, César Augusto Aguirre and Armando B. Brizuela

Abstract

In this chapter, an Euler-Lagrangian double-way coupled model is presented for simulating the liquid particle dispersion ejected from a high-pressure nozzle. The Eulerian code is advanced regional prediction system (ARPS), developed by Center of Analysis and Prediction of Storm (CAPS) and Oklahoma University, USA, which is specialized in weather simulation. This code is the double way coupled with a Lagrangian one-particle model. The theoretical remarks of the double-way coupling, the simulation of the liquid droplet trajectory, and, finally, the droplet collision in the spray cloud using a binary collision model are descripts. The results of droplet velocities and diameters are compared with experimental laboratory measurements. Finally, agrochemical spraying over a cultivated field in weak wind and high air temperature conditions is showed.

Keywords: droplets, spray, multiphase flow, large-eddy simulation, Lagrangian stochastic model

1. Introduction

Numerous engineering applications are focused on solving the problems of dispersion of a sprayed droplet jet from a high-pressure nozzle into a gaseous medium. Spraying is used in internal combustion engines, application of agrochemicals over cultivated fields and greenhouses, irrigation systems, among others. Some questions raised by this topic of engineering can be studied using the computational simulation of multiphase flows. There are currently different ways to implement these tools. In the Eulerian approach, the physical domain is subdivided into cells of a grid space. Each cell has a portion of its volume filled by the liquid phase and the other part by the air. Continuity, momentum, energy and species, for a single-fluid mixture conservation equations, are solved in all pass time of the simulation [1, 2]. On the other hand, the Lagrangian approach at one particle proposes the velocities and positions of particles simulation (solids, liquid, vapor, or scalar species) by solving a stochastic equation following the Markov chains. The deterministic term is obtained from the average air velocity values, while the random term is like a white noise following a Brownian motion. The coupled Eulerian large-eddy simulation (LES) with Lagrangian one-particle stochastic method (STO) has been proposed in order to obtain more details on the turbulent properties of the fluid carrying the particles. Several studies using this coupling methodology (LES-STO) can be found [3–11]. In this chapter, we focus on the ejection of droplets in air environment from a spray nozzle. The sprayed liquid is a water at 20°C temperature, and the ejection pressure reaches 3 bar. The atmosphere temperature is like the water ejected, but the air pressure is 1.013 bar. These conditions are like as Nuyttens experience [12]. The author carries out paired measurements of droplet diameters and velocities at 25 cm below the spray nozzle using phase Doppler particle analyzer (PDPA) instrument. The particle's Euler-Lagrangian double-way coupling code LES-STO is proposed for to simulate the trajectory of these particles in their liquid phase. The original finite-difference Eulerian LES code named advanced regional prediction systems (ARPS) developed by the University of Oklahoma's Center for Analysis and Forecasting of Storms (CAPS) [13] has been adapted by Aguirre [14] for the simulation of fluid particles in order to validate it with measurements of concentration of a passive gas made in a wind tunnel over flat ground [15] and in the presence of a gentle sloping hill [16]. First time, we present a random ejection algorithm of droplet diameters whose probability density function replies to the two-parameter Weibull distribution. These parameters are previously obtained using laboratory experimental data. Second time, we present the theoretical approach for obtaining the results of collision droplets into the spray. The binary collision droplet model [17–19] has been performed in the LES-STO code. This model uses the concept of symmetric weber number [20] to consider the relationship between the kinetic and surface energy of the two colliding droplets. Finally, an agrochemical spraying over a cultivated field in low wind velocity and high air temperature conditions is showed.

2. Theoretical framework and techniques of numerical simulation

2.1 Conditions and simulation of liquid particle ejection

In this section, we present a random ejection algorithm for simulating different diameters of droplets whose probability density function matches a Weibull distribution. The scale and shape parameters of Weibull distribution are previously obtained from laboratory experimental data using a phase Doppler particle analyzer (PDPA) performed by Nuyttens [12] from an HARDI™ spray nozzle. The sprayed liquid in laboratory experience has been water at 20°C temperature, and the ejection pressure reaches 3 bar. The atmosphere temperature is like the water ejected, but the air pressure is 1.013 bar and a calm wind.

2.1.1 Initial conditions of the droplet positions and velocities

The initial conditions of ejection droplets are as follows:

- The nozzle height was located at h = 0.75 m over the ground.
- The elliptical shape A (**Figure 1**) for exit droplets. It was located at L_c = 0.023 m below the nozzle because, according to the measurements of Nuyttens [12], it is the region of detachment of droplets from the sheet of the liquid film.
- The angle of the spray in the transverse direction is α = 110° according to the technical specifications of the HARDI™ nozzle.

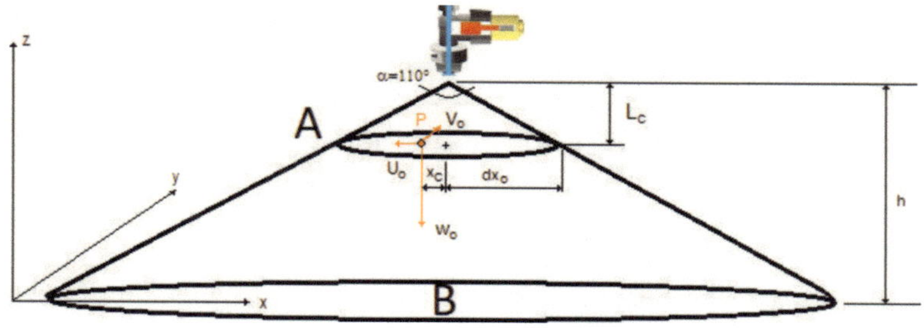

Figure 1.
Spray cone HARDI ™ ISO F 110-03 nozzle follows Nuyttens laboratory experience [12].

- The minor semiaxis of A ellipse (transverse direction y) is dy_O = 0.0046 m.
- The major semiaxis of A ellipse (longitudinal direction x) is $dx_O = tg\,(\alpha/2)L_c = 0.0328$ m.
- The initial vertical velocity of liquid particles is adopted from the confined fluid simulation [21] W_O = 16.356 m s^{-1}.
- The initial horizontal component velocity of each liquid particle will depend on the initial position within the A ellipse (**Figure 1**):

$$\begin{cases} U_O = \dfrac{x_c}{L_c} W_O, \\ V_O = \dfrac{y_c}{L_c} W_O, \end{cases} \tag{1}$$

where (x_c, y_c) are the relative horizontal positions of the liquid particle at center of the A ellipse. So, the horizontal velocity of droplets at initial time of the simulation depends directly on these relative distances. To determine the initial horizontal position (x_c, y_c), an algorithm of random variable is used:

$$\begin{cases} x_c = 2dx_o(\chi - 0.5), \\ y_c = 2\left(\left(dx_o^2 - x_c^2\right)\left(\frac{dx_o^2}{dy_o^2}\right)\right)^{1/2}(\chi - 0.5), \end{cases} \tag{2}$$

where χ is a continuous uniform random variable in the [0, 1] interval whose average value is $\mu_\chi = 0.5$ and standard deviation $\sigma_\chi = \sqrt{3}/6$.

All particles are located at the $z = h - L_c$ height within the A ellipse at the initial time of the simulation.

2.1.2 Initial distribution function of the liquid particles' diameters

The Rosin-Rammler (R-R) distribution function is a cumulative function of continuous random variable whose probability density function (p.d.f.) is a two-parameter Weibull. This distribution function is used [22, 23] to adjust experimental data of droplet diameter measurements as a function of liquid-sprayed fraction volume in order to obtain the shape m and scale k Weibull parameters. The experimental data require very precise measurements of the diameters of liquid droplets. The Doppler phase particle analyzer (PDPA) meets the necessary requirements and has the advantage of obtaining paired velocity and diameter data of very small droplets, which are ejected from the nozzle. Nuyttens [12] presents laboratory measurements with this device for different working pressures with an HARDI ™ ISO F 110-O3 nozzle, among others, in calm air conditions. The values obtained in the Nuyttens [12] measurement experiences are as follows: average droplet diameters $\overline{\phi_0} = 267.6\ \mu m$, standard deviation of droplet diameters $\sigma_\phi = 110.3$ μm, and a modal value $Mo = 250.2\ \mu m$. These values allow finding the parameters of the Weibull p.d.f. for the initial conditions of ejected droplet diameter. In this work, we obtained the m and k parameters following the methodology presented [24] for the droplet diameters that are ejected with an internal pressure of 3 bar. The properties of the Weibull distribution and the fit with the experimental measurement data are described in Appendix A.

2.1.3 Initial randomization of the droplet diameters

Once the shape and scale parameters of the Weibull p.d.f. are obtained, which characterize the diameters of drops ejected from the spray nozzle, it is necessary to carry out a temporal sequence for the simulation of these diameters. An algorithm based on the function of the random variable χ, already used, is proposed for the initial position of droplets in the A ellipse (**Figure 1**). The randomization algorithm should allow the diameters to be assigned to each ejected droplets such that the mean and standard deviation of p.d.f. droplets simulated over a long period time are close to those corresponding to the Weibull p.d.f. Using the central limit theorem for a set of values corresponding to the random variable χ, a normalized random variable Z of mean value $\overline{Z} = 0$ and standard deviation $\sigma_Z = 1$ can be obtained

$$Z = \frac{\overline{\chi} - \mu_\chi}{\frac{\sigma_\chi}{\sqrt{n}}}, \tag{3}$$

where n is the sample size. Michelot [25] performed several tests with different n values using 1 million particles to find an acceptable number from computational cost time. The author concludes that n = 50 is a good value to obtain the standard normal random variable for generating random numbers with χ whose $\mu_\chi = 0.5$ and $\sigma_\chi = \sqrt{3}/6$.

With this method, it is possible to simulate diameters of liquid particles that follow the normal distribution from random number generation

$$\phi_0 = \overline{\phi_0} + \sigma_\phi Z, \tag{4}$$

where $\overline{\phi_0}$ and σ_ϕ are the mean and standard deviation values of the Weibull p.d.f., respectively, whose expressions are given as:

$$\begin{cases} \overline{\phi_0} = k.\Gamma\left(1 + \frac{1}{m}\right), \\ \sigma_\phi = k.\left\{\Gamma\left(1 + \frac{2}{m}\right) - \left[\Gamma\left(1 + \frac{1}{m}\right)\right]^2\right\}^{1/2} \end{cases}. \tag{5}$$

However, in Eq. (4), we do not consider the asymmetry of the Weibull p.d.f. For this, it is necessary to incorporate the mode (Mo) of Weibull p.d.f. whose expression is given as:

$$Mo = k\left(\frac{m-1}{m}\right)^{1/m}. \tag{6}$$

Incorporating this value into Eq. (4), the random variable ϕ can be written as:

$$\begin{cases} \phi_0 = Mo + \sigma_\phi . Z, & \overline{\phi_0} = Mo, \quad \text{symmetric (normal p.d.f.)}, \\ \phi_0 = Mo + \sigma_\phi . Z . \exp\left(\dfrac{Z}{2(\overline{\phi_0} - Mo)}\right), & \overline{\phi_0} \neq Mo, \quad \text{asymmetric (Weibull p.d.f.)}. \end{cases} \tag{7}$$

Eqs. (1), (2), and (7) provide the initial conditions of velocities, positions, and diameters of the ejected liquid particles from the HARDI™ ISO F110-O3 nozzle.

2.1.4 Dynamic parameters of liquid particles

Several simplifications are imposed at ejection and trajectory simulation of liquid particle phenomena:

- Particles are considered to have a constant spherical shape in their trajectory.
- The rotating motion of the particle is not considered.
- The ratio between droplet and air densities is very large.

Assuming these simplifications, the force per unit mass to which the liquid particles are submitted is based on a balance between gravity and drag forces per unit mass:

$$\frac{F_{l\ i}}{m_l} = \underbrace{\frac{U_i - V_i}{\tau}}_{\substack{\text{Drag} \\ \text{acceleration}}} - \underbrace{g_i \delta_{i3}}_{\substack{\text{Gravity} \\ \text{acceleration}}}, \tag{8}$$

where F_{li} is the force actuating over l liquid particle in i direction i = 1, 2, 3 (x, y, z) on Cartesian coordinate system (**Figure 1**), m_l is the mass of liquid particle, U_i is the air velocity, V_i is the liquid particle velocity, and τ is the characteristic time response or relaxation time of liquid particle, which represents the time required for the liquid adapt to sudden changes in air velocity. This last parameter can be estimated [26] as:

$$\tau = \frac{4}{3}\frac{\varphi}{C_D}\frac{\rho_l}{\rho}\frac{1}{|U_i - V_i|}, \tag{9}$$

where C_D is the dynamic drag coefficient due to the air viscosity, ρ_l is the liquid particle density, and ρ is the air density. It is important to note that, if the spray injection pressure is increased, the relative velocity between the droplets and the air will be higher. This implies that the relaxation time will be decreased. In addition, if the diameter of the liquid particle decreases, the relaxation time will also be shorter. As the droplet is considered spherical, the drag coefficient C_D depends on both, the diameter of droplet and the air viscosity, which is used for the calculation of the Reynolds number referred to the droplet $\mathfrak{Re}_l$:

$$\mathfrak{Re}_l = \phi \frac{|U_i - V_i|}{\nu}. \tag{10}$$

Several drag coefficient expressions have been analyzed [27]. The authors showed that Turton expression [28] has given better results in this simulation case:

$$\begin{cases} C_D = \dfrac{24}{\mathfrak{Re}_l}\left(1 + 0.173\mathfrak{Re}_l^{0.657}\right) + \dfrac{0.413}{1 + 16300\mathfrak{Re}_l^{-1.09}}, & \text{if } \mathfrak{Re}_l < 2 \times 10^5 \\ C_D = 0.465, & \text{if } \mathfrak{Re}_l \geq 2 \times 10^5 \end{cases}. \quad (11)$$

When the drag and gravity forces are balanced, the liquid particles reach the sedimentation regime. In this case of free fall, the droplets have only vertical velocity component. This velocity is named sedimentation velocity $|U_i - V_i|\delta_{i3} = V_s$. From Eqs. (8) and (9):

$$V_s = \left(\frac{4}{3}\frac{\rho_l \phi g_i \delta_{i3}}{\rho C_{D,s}}\right)^{1/2}. \quad (12)$$

Note that $C_{D,s}$ is the drag coefficient at sedimentation regime. It depends on $\mathfrak{Re}_l$ (Eq. (11)) and therefore on the sedimentation velocity itself (Eq. (10)). So, V_s can only be calculated iteratively.

The time elapsed until the particle reaches the sedimentation velocity can be written as:

$$\tau_s = \frac{V_s}{g}. \quad (13)$$

This is an important parameter of liquid particles because if the time elapsed until the liquid particle reaches the ground is longer than the sedimentation time, it will be exposed to drift.

2.2 Euler-Lagrangian double-way coupled model

The double-way coupled model presents a bidirectional coupling between the Eulerian and Lagrangian equation systems. Based on the Eulerian approach, the large-eddy simulation (LES) technique is proposed to obtain a detailed turbulent flow. The turbulent intensity of the fluid that transports the liquid particles is taken into account in the simulation of its trajectories. In this approach, it is not possible to obtain a full description of all eddies, so the LES technique is applied for resolving the larges scales of turbulence. The small scales are modeled by subgrid eddy viscosity model (SGS). A dynamic SGS model proposed by Germano [29] is implemented in ARPS by Aguirre [14]. On the other hand, the Lagrangian form is proposed to simulate the trajectories of the liquid particles. In the double-way-coupled LES-STO model, it is considered that the intensity of turbulent flow is taken into account in Lagrangian stochastic equation, and the presence of the liquid particles is taken into account in the momentum equation for LES.

2.2.1 The Lagrangian stochastic model

The governing equations of the liquid particles trajectories are based on a Lagrangian stochastic model at a one-particle and one-time scale following the classical equation of Langevin. The air velocity model at liquid particle position U_i has a deterministic term and a random term:

$$\frac{dU_i}{dt} = \underbrace{h_{ij}(U_i,t)}_{\text{Deterministic}} + \underbrace{q_{ij}(U_i,t)\eta_j(t)}_{\text{Random}}. \quad (14)$$

The tensors $h_{ij}(U_i, t)$ and $q_{ij}(U_i, t)$ are determined dynamically according to the characteristics of the turbulence at each position of the simulation and at each instant time. This requires that the LES equations are coupling with Lagrangian stochastic model. In Eq. (14), $\eta(t)$ denotes the random characteristic variable of zero mean and covariance:

$$\left\langle \eta_i(t')\eta_j(t'') \right\rangle = \delta_{ij}\delta(t' - t''). \tag{15}$$

2.2.2 The Eulerian flow model

It is necessary to simulate the air velocity at liquid particle position U_i. The LES method decomposes in a resolved component $u_i^{\oplus}$ and a fluctuation u_i^{-}:

$$U_i = u_i^{\oplus} + u_i^{-}. \tag{16}$$

The LES code advanced regional prediction system (ARPS) developed by Center of Analysis and Prediction of Storm (CAPS) and Oklahoma University [13] numerically integrates the time-dependent equations of mass balance, forces and energy of the largest turbulent scales. Filtered continuity as in Eq. (17), filtered momentum of fluid velocity as in Eq. (18), and filtered momentum of scalars as in Eq. (19) are described as follows:

$$\frac{\partial \tilde{u}_i^{\oplus}}{\partial x_i} = 0, \tag{17}$$

$$\frac{\partial \tilde{u}_i^{\oplus}}{\partial t} + \frac{\partial\left(\tilde{u}_i^{\oplus} u_j^{\oplus}\right)}{\partial x_j} = \bar{\rho} g_i B^{\oplus} - \frac{\partial p\prime^{\oplus}}{\partial x_i} - \frac{\partial \tilde{\tau}_{ij}}{\partial x_j} + 2\nu \frac{\partial \tilde{S}_{ij}^{a\oplus}}{\partial x_j} - \Im_i, \tag{18}$$

$$\frac{\partial \tilde{\psi}^{\oplus}}{\partial t} + \frac{\partial\left(\tilde{u}_j^{\oplus} \psi^{\oplus}\right)}{\partial x_j} = \Phi_{\psi} - \frac{\partial \tilde{\tau}_{i\psi}}{\partial x_j}, \tag{19}$$

where $B^{\oplus}$ is a buoyancy force, $\tilde{S}_{ij}^{a}$ is the anisotropic deformation tensor, ν is the molecular viscosity, and Φ_{ψ} represents the sink and sources of the scalars variables ψ. The variables with tilde indicate that they have been weighted by the density of air $\left(\tilde{u}_i^{\oplus} = \bar{\rho} u_i^{\oplus}\right)$, which is only dependent of z (vertical) height. The pressure equation is obtained using the material derivative of the state equation for moist air and replacing the time derivative of density by velocity divergence using the continuity equation. The correlation terms containing unsolved scales $\tilde{\tau}_{ij}$ and $\tilde{\tau}_{i\psi}$ are modeled using the dynamic Smagorinsky formulation [29].

2.2.3 Lagrangian to the Eulerian coupling

From Lagrangian stochastic equation to the Eulerian LES model taken into account, the number of liquid particles is very large near the nozzle. The coupling has been computed by adding $\Im_i$ term at filtered momentum of fluid velocity Eq. (18), which expresses the additional momentum due to the presence of liquid particles per volume of carrier fluid:

$$\Im_i = \frac{1}{\Delta V} \sum_{l=1}^{n_{\Delta}} F_{li} \tag{20}$$

where $F_{l\,i}$ is the force of the liquid particle l in i direction, $\Delta V = \Delta x \Delta y \Delta z$ is the grid cell volume, and n_Δ is the number of liquid particles within the grid cell. Using Eq. (8), the additional momentum is given by:

$$\mathfrak{I}_i = \frac{1}{\Delta V} \sum_{l=1}^{n_\Delta} \left[m_l \left(\frac{U_i - V_i}{\tau} - g_j \delta_{j3} \right) \right]. \tag{21}$$

2.2.4 Eulerian to Lagrangian coupling

Aguirre and Brizuela [11] show that the coupling LES-STO model allows to find the expressions of the deterministic and random terms of Eq. (14) using the velocity-filtered density function (VFDF) proposed by Gicquel et al. [30]:

$$\begin{cases} h_{ij}(U_i, t) = \dfrac{du_j^{\oplus}}{dt} + \alpha_{ij} u_j^{-}, \\ q_{ij}(U_i, t) = \sqrt{C_0 \varepsilon}\, \delta_{ij}. \end{cases} \tag{22}$$

The material derivatives of the velocity-filtered air flow, subgrid turbulent kinetic energy K^{-}, and energy molecular dissipation ε are calculated using the ARPS code at each position and time step of the simulation. The Kolmogorov constant value is $C_O = 2.1$. Therefore, we only need to evaluate the α_{ij} tensor. Aguirre and Brizuela [11] propose the expression of α_{ij} for inhomogeneous and anisotropic turbulence:

$$\alpha_{ij} = \frac{1}{2K^{-}} \frac{dK^{-}}{dt} \delta_{ij} - \left(\frac{3}{4} C_0 \right) \frac{\varepsilon}{K^{-}} \delta_{ij} + \left(\frac{R_{ij}}{2K^{-}} - \frac{\delta_{ij}}{3} \right) \frac{\varepsilon}{K^{-}}. \tag{23}$$

The subgrid turbulent kinetic energy is solved by 1.5 order transport equation [31] and $R_{ij} = \left(u_i^{-} u_j^{-} \right)^{\oplus}$ is the Reynolds SGS stress tensor.

The unresolved velocity component u_j^{-} in Eq. (16) is obtained in discrete form using the Markov chains:

$$u_{j\,(n)}^{-} = u_{j\,(n-1)}^{-} + \alpha_{ij\,(n)} u_{j\,(n-1)}^{-} \Delta t + \sqrt{C_0 \varepsilon \Delta t}\, \chi_{(n)}, \tag{24}$$

where the subscript (n) denotes the value at present time of the simulation, while $(n-1)$ is the value in the previous time step. The first pass time, subscript (0), is considered as isotropic homogeneous turbulence [32]:

$$u_{j(0)}^{-} = \sqrt{\frac{2}{3} K_{(0)}^{-}}\, \chi_{(0)}. \tag{25}$$

With Eqs. (16) and (22–25), it is possible to calculate the air velocity at the liquid particle position. The equations describing the motion of the liquid particle in its discrete form are:

$$\begin{cases} V_{i(n+1)} = V_{i(n)} + \dfrac{\Delta t}{\tau} \left(U_{i(n)} - V_{i(n)} \right) - g_i \Delta t \delta_{i3}, & \text{if } \tau > \Delta t, \\ V_{i(n+1)} = U_{i(n)} - V_s \delta_{i3}, & \text{if } \tau \leq \Delta t, \end{cases}$$
$$\left\{ X_{i(n+1)} = \frac{V_{i(n+1)} - V_{i(n)}}{2} \Delta t. \right. \tag{26}$$

It is necessary to note that in the first Eq. (26), the rate $\Delta t/\tau$ must be greater than one for convergence of the numerical solution. Cases in which the liquid particle diameter is very small, the second Eq. (26) must be used considering that the sedimentation velocity Vs has been reached before Δt time step has elapsed. For this reason, the simulation time step Δt must be chosen less than the relaxation time of the smallest possible liquid particle. According to the experimental measurements by Nuyttens [12], the liquid particles whose diameters are smaller to 50 μm are exposed to drift before reaching the ground. These particles have relaxation times less than 7.6 ms. So, a simulation time step Δt = 0.2 ms has been chosen due to little size of cell grid ΔV, whereby liquid particles whose diameters are smaller than 7 μm are being considered for calculation with the second Eq. (26).

2.3 Binary collision droplet model

Once the droplets ejected from the spray nozzle and having simulated their positions, velocities and diameters along their trajectory, it is necessary to consider the collision. Binary droplet collision models are a widely used theoretical approximation to obtain the outcome of the interaction droplets [17–20, 33–39]. This model consists of estimating the positions, velocities, and diameters of droplets after the collision. In addition, satellite droplets can be created from the ligament breakup as a consequence of it.

2.3.1 Parameters of binary collision

The binary droplet collision is simulated using three important parameters. The ratio of the droplet diameters Δ (Eq. (27)), the dimensionless symmetric Weber number (Wes) [20] relating kinetic energy vs. surface energy (Eq. (28)), and the dimensionless impact parameter (Imp) takes into account the way in which the two droplets impact (Eq. (29)):

$$\Delta = \frac{\phi_S}{\phi_L}, \tag{27}$$

$$\text{Wes} = \frac{\rho_l \phi_S \Delta^3 \left|\vec{V}_{mS}\right|^2 + \left|\vec{V}_{mL}\right|^2}{12\sigma\Delta\left(1+\Delta^2\right)}, \tag{28}$$

$$\text{Imp} = \frac{2X}{\phi_L + \phi_S}, \tag{29}$$

where the subscripts S and L indicate the smaller and larger droplet, respectively, σ denote the surface tension, $\vec{V}_{mS}$ and $\vec{V}_{mL}$ are the relative velocities to the mass center of the incoming droplets, and X is the projection of the distance between the droplet centers in the normal direction to the relative velocity $\vec{V}_R = \vec{V}_S - \vec{V}_L$ as shown in **Figure 2**.

$$\begin{cases} \vec{V}_{mL} = \vec{V}_L - \vec{V}_{mR}, \\ \vec{V}_{mS} = \vec{V}_S - \vec{V}_{mR} \end{cases}, \tag{30}$$

In Eq. (30), $\vec{V}_{mR}$ is the velocity of mass center. If the droplets have the same density:

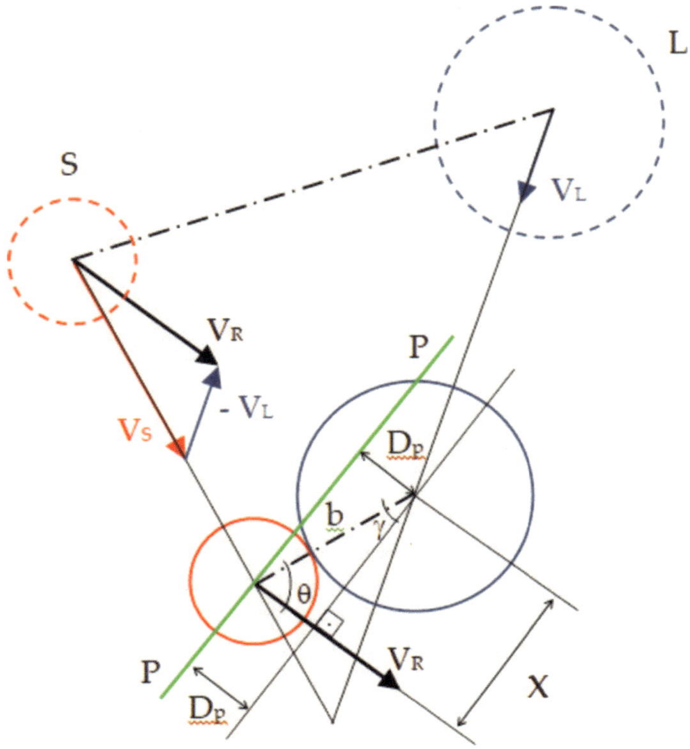

Figure 2.
Scheme of small droplet S and large droplet L before collision (dashed line) and at contact instant (solid line).

$$\vec{V}_{mR} = \frac{\vec{V}_S \phi_S^3 + \vec{V}_L \phi_L^3}{\phi_S^3 + \phi_L^3} \tag{31}$$

The relative velocity droplets of the mass center can be resumed using Eq. (27):

$$\begin{cases} \vec{V}_{mL} = +\dfrac{\Delta^3 \vec{V}_{mR}}{\Delta^3 + 1}, \\ \vec{V}_{mS} = -\dfrac{\vec{V}_{mR}}{\Delta^3 + 1}. \end{cases} \tag{32}$$

For the impact parameter in Eq. (29), it is necessary to compute X variable. This variable is the projection of segment $b = 0.5(\phi_S + \phi_L)$ on the plane perpendicular to the relative velocity $\vec{V}_R$. The impact factor will be equal to the cosine of γ angle:

$$\text{Imp} = \cos\gamma = \frac{X}{b}, \tag{33}$$

$$\sin\gamma = \frac{D_p}{b}, \tag{34}$$

So, inserting Eq. (34) into Eq. (33) results in:

$$\text{Imp} = \cos\gamma = \left[1 - \left(\frac{2D_p}{\phi_S + \phi_L}\right)^2\right]^{\frac{1}{2}}. \tag{35}$$

It is necessary to obtain D_p. It is the distance from large droplet center (x_L, y_L, z_L) to the perpendicular plane at $\vec{V}_R$ velocity passing through the small droplet center (x_S, y_S, z_S). This plane P-P is shown in green color in **Figure 2**. The plane equation passing through the small drop center is $u_R x_S + v_R y_S + w_R z_S + D = 0$, where (u_R, v_R, w_R) are the components of $\vec{V}_R$. In this expression, D is a constant of plane equation. This constant is obtained as: $D = -u_R x_S - v_R y_S - w_R z_R$. So, the D_p distance can be obtained as:

$$D_p = \left| \frac{u_R x_L + v_R y_L + w_R z_L + D}{\left(u_R{}^2 + v_R{}^2 + w_R{}^2\right)^{\frac{1}{2}}} \right|. \tag{36}$$

2.3.2 Numerical resolution of the impact coefficient

In each time of numerical simulation, it checks whether the collision between two droplets occurs. For obtaining a more optimize algorithm, collision boxes are placed around and inside the liquid particle ejection spray. The sizes of grid boxes vary dynamically, adjusting to the boundaries of the particle domain as shown in **Figure 3a**. The size of the boxes is the same as the Eulerian calculation grid in horizontal direction $\Delta b = \Delta x = \Delta y = 0.1$ m. In this way, every drop inside this box will be questioned about whether it collided with the other drops that are in the same box. When a binary collision is successfully found (e.g., 5–6; 3–7 in **Figure 3b**), the pairs are marked and removed from the next iteration of detection. This technique was proposed by Michelot [25] and used by Aguirre [4, 14] to consider the diffusion of chemical species that are carried by fluid particles. It is evident that due to the temporal discretization of the numerical solution of droplet motion equations, it is almost impossible that for an instant of discretized time $t_{(n)} = t_{(n-1)} + \Delta t$, the contact between two drops can be concurrent. Most likely, by that instant time $t_{(n-1)}$, the contact is about to occur and at instant time $t_{(n)}$, it has already occurred. When the distance between the centers of the two drops inside a collision box is less than the sum of their radii, then the drops collided. In this case, the time $\Delta t''$ elapsed from the collision to the computation instant time $t_{(n)}$ to be calculated. The particles are repositioned at the moment of collision $t'' = t_{(n)} - \Delta t''$, and the impact factor is calculated according to the positions of their centers as shown in **Figure 4**. The lapse time $\Delta t''$ for repositioning the droplets at instant of collision is computed in resolving:

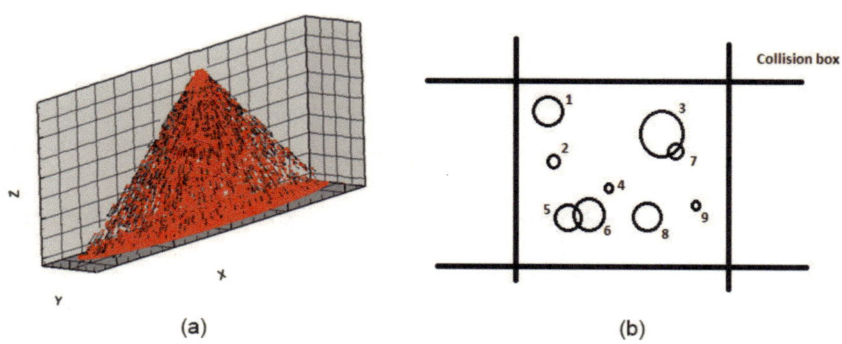

Figure 3.
(a) Collision boxes around and inside the spray ejection of droplets and (b) droplets inside the collision box at $t_{(n)}$ *instant time of the simulation.*

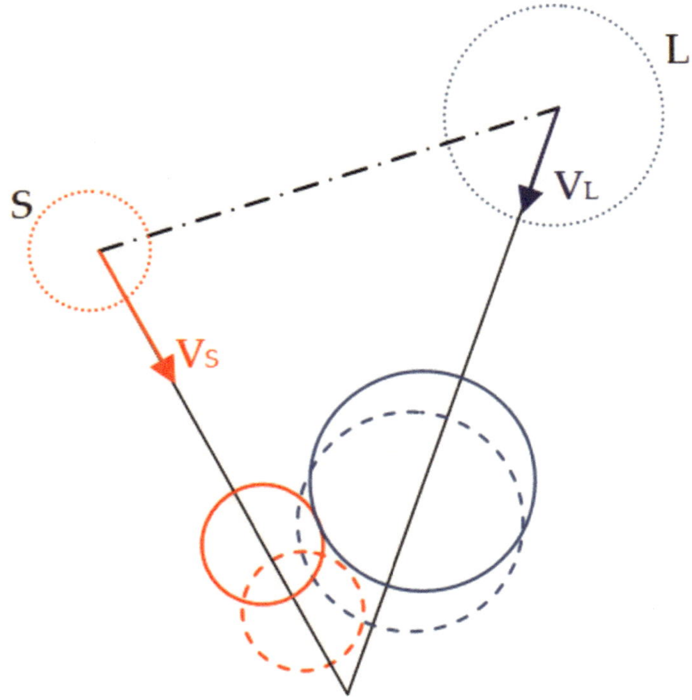

Figure 4.
Positions of droplets before and after the collision for a time lapse Δt.··········· *Droplets before collision at* $t_{(n-1)}$ *instant time.* - - - - - *Droplets after collision at* $t_{(n)} = t_{(n-1)} + \Delta t$ *instant time.* ——— *Repositioning of droplets at the instant of collision* ($t'' = t_{(n)} - \Delta t''$).

$$\left(\frac{\phi_S + \phi_L}{2}\right)^2 = \left(x_S - x_L - u_R \Delta t''\right)^2 + \left(y_S - y_L - v_R \Delta t''\right)^2 + \left(z_S - z_L - w_R \Delta t''\right)^2. \tag{37}$$

The outcomes of collision droplets are computed using the map collision theory. Once the droplets collided and the effects of collision are into account on the droplets, they are repositioned by advancing the same pass time $\Delta t''$.

2.3.3 Binary droplet collision map

The outcomes of the binary droplet collision model propose different scenarios:

[C] Coalescence: the two droplets that collide to form a single drop as a result of the collision. In this case, the surface energy is relatively greater than the kinetic energy.

[R] Reflexive: the two colliding droplets almost head-on, so they join together as one, but the kinetic energy is large enough to separate again and can generate satellite droplets.

[S] Stretching: the two drops collide tangentially, so they separate and can generate satellite droplets.

[B] Bouncing: the two colliding drops remain separated after collision without exchanging mass between them.

Figure 5 shows a time sequence of the binary droplet collision for each outcome described above. It is important to note that the result of the binary collision depends

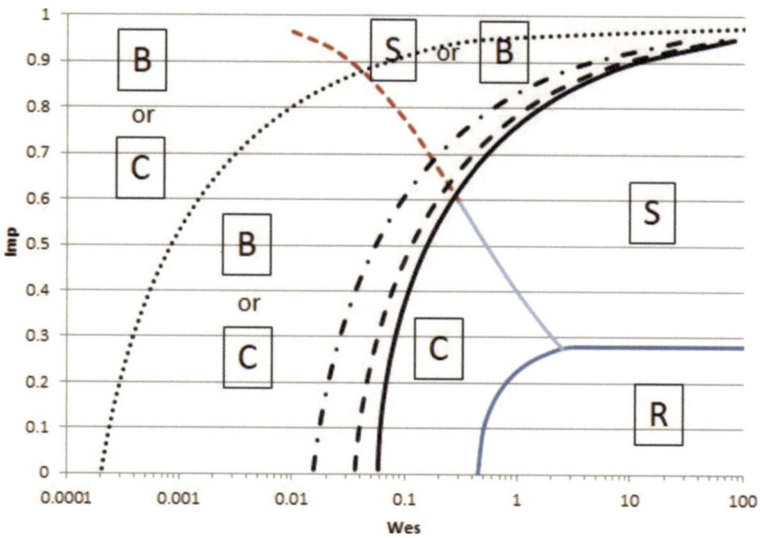

Figure 5.
Time sequence diagram of the binary droplet collision and its outcomes.

not only on the velocity of both drops but also on their relative size and impact coefficient. Two of the four possible outcomes of the binary collision are susceptible to generating satellite droplets. These droplets are usually much smaller in size than the parent-drops and are, therefore, more prone to drift and evaporation. If these droplets are composed of a phosphonate-acid solution (such as glyphosate), then after evaporation, the solute will drift away from the airflow very quickly.

The outcomes of collision droplets are defined using a map collision. This map is the graphic representations between the Wes vs. Imp (Wes-Imp) frontier curves among the different outcomes of binary collision that are displayed on this map. Several researchers proposed equations for frontier curves. The transition impact factor between coalescence and stretching separation ($Imp_{c\text{-}s}$) is according to Rabe [20] as follows:

$$Imp_{c\text{-}s} = \frac{\sqrt{0.53^2 + 4.24Wes} - 0.53}{4Wes}. \tag{38}$$

The transition impact factor between coalescence and reflexive separation ($Imp_{c\text{-}r}$) is:

$$Imp_{c\text{-}r} = 0.3059\sqrt{1 - \frac{0.45}{Wes}}. \tag{39}$$

The transition impact factor between reflexive and stretching separation appears when the Wes > 2.5 and can be considered a constant value $Imp_{r\text{-}s} = 0.28$.

It should be noted that the boundary curve between coalescence and reflexive separation $Imp_{c\text{-}r}$ increases with the increase of Wes to the value of Imp = 0.28. This behavior indicates that for low Imp values (on-head collision) and relatively low droplet velocities before collision, surface energy is greater than kinetic energy and the result of the collision is stable coalescence. However, for the same Imp values but with higher velocities, the kinetic energy is predominant; the droplets have an unstable coalescence and then separate. This separation can generate satellite droplets. On the other hand, if the Imp is higher (tangential collision of the droplets), then coalescence as a result of the collision is more improbable since only a fraction of the volume of the drops interacts during the collision. The contact surface of both drops is smaller and therefore the surface energy as well. This reduces the likelihood of stable coalescence as a consequence of the collision. This behavior is evident in the $Imp_{c\text{-}s}$ frontier curve, which decreases the coalescence area as the Imp increases.

For bounce, the model proposed by Estrade [35] calculates the number of transition Weber We_b according to the Imp, Δ and a shape parameter, φ:

$$We_b = \frac{\Delta\left(1+\Delta^2\right)\left(4\varphi - 12\right)}{\xi\left(1-\text{Imp}^2\right)}, \tag{40}$$

where ξ is computed as:

$$\xi = \begin{cases} 1 - \dfrac{(2-\lambda)^2(1+\lambda)}{4} & \text{if } \lambda > 1, \\ \dfrac{\lambda^2(3-\lambda)}{4} & \text{if } \lambda \le 1, \end{cases} \tag{41}$$

and $\lambda = (1-\text{Imp})(1+\Delta)$. The shape parameter φ can be computed as Zhang [19]:

$$\varphi = 3.351\left(\frac{\rho_l}{1.16}\right)^{\frac{2}{3}}. \tag{42}$$

The transition bounces into Wes-Imp map collision droplets, and the Weber symmetric bounce frontier Wes_b is used. So, it is obtained from We_b (Eq. (40)) as:

$$Wes_b = We_b \frac{\Delta^2}{12\left(1+\Delta^3\right)\left(1+\Delta^2\right)}, \tag{43}$$

The Wes-Imp map collision droplets define areas C, R, S and B where the outcomes of the binary droplet collision are represented. These areas are bounded by frontiers curves as proposed in Eqs. (38)–(43). The areas with frontier curves are shown in **Figure 6**. The frontier curves between C - B and S - B change their position as a function of Δ.

2.3.4 Numerical models of the binary droplet collision

The binary droplet collision model allows obtaining the diameters and velocities of the droplets after the collision. The values of these variables are obtained according to the proposed models [17–19, 34] (coalescence, reflexive, and stretching separations) and [35] (bounce outcome).

2.3.4.1 Coalescence

For coalescence outcome, the two droplets coalesce into one. This occurs preferably at low Weber numbers as surface tensions exceed kinetic energy. The new droplet velocity is the velocity of mass center before the collision $\vec{V}_{(new)} = \vec{V}_{mR}$ (Eq. (31)). For droplets of the same density, its diameter is $\phi_{(new)} = \left(\phi_L^3 + \phi_S^3\right)^{1/3}$.

2.3.4.2 Stretching

Munnannur and Reitz [17] calculate the interaction volume between the droplets. This volume is released from both drops creating a ligament that gives rise

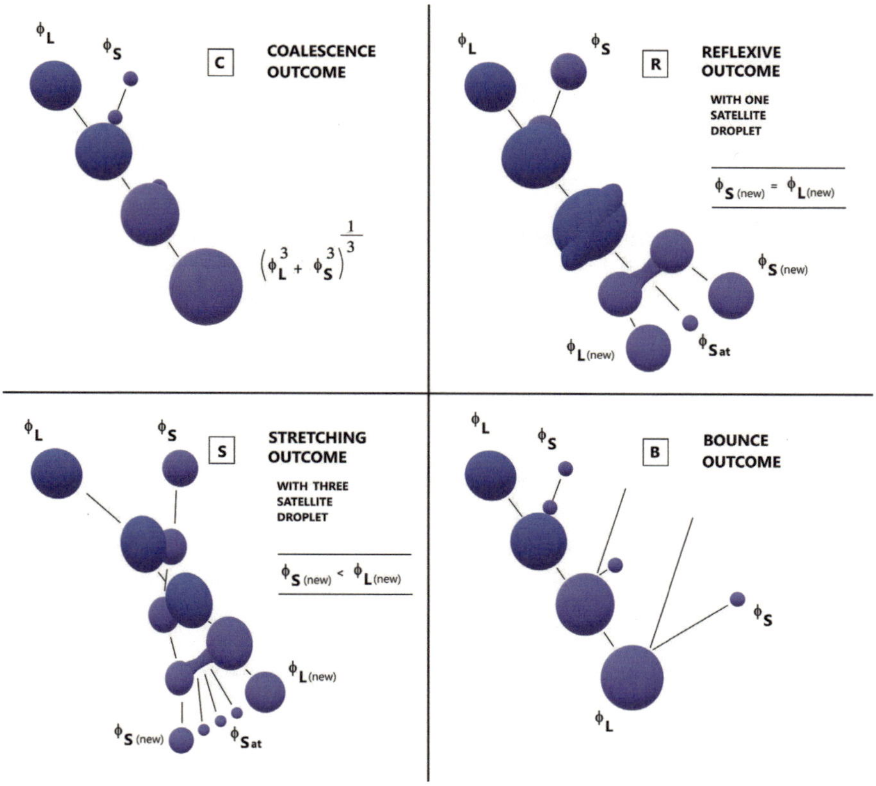

Figure 6.
Map collision droplets with areas of outcomes collision. C *Coalescence,* R *reflexive separation,* S *stretching separation,* B *and bounce. Bounce frontiers:* ——— Δ = 1, - - - - · Δ = 0.75, - · - · - · Δ = 0.5, ············ *and* Δ = 0.1.

(or not) to satellite droplets. This volume is computed and taken into account the magnitude of the opposing surface (E_{surten}), stretching (E_{strtch}), and viscous dissipation energies (E_{dissip}) by using a separation coefficient (C_{VS}):

$$C_{VS} = \frac{E_{strtch} - E_{surten} - E_{dissip}}{E_{strtch} + E_{surten} + E_{dissip}}, \tag{44}$$

Case $C_{VS} \leq 0$**:** if this coefficient is carried at negative value, it is assumed that fragmentation of droplets does not occur. The drops only lose kinetic energy. The center mass of droplet velocities is affected by Z coefficient indicating the fraction of energy that is dissipated during collision. For this case, Kim [18] proposes:

$$Z = \frac{\text{Imp} - \sqrt{e_{coal}}}{1 - \sqrt{e_{coal}}}, \tag{45}$$

where $e_{coal} = \min\left[1.0, 2.4f\ \text{We}^{-1}\right]$ is the coalescence efficiency, $f = \Delta^{-3} - 2.4\Delta^{-2} + 2.7\Delta^{-1}$, and $\text{We} = \rho_l \phi_S \Delta^3 \sigma^{-1} \left|\vec{V}_R\right|^2$, is the Weber number, which follow O'Rourke model [40].

The relative velocities of mass center after collision can be written by using momentum conservation equation:

$$\begin{cases} \vec{V}_{mS\ (new)} = Z\vec{V}_{mS} \\ \vec{V}_{mL\ (new)} = Z\vec{V}_{mL} \end{cases}. \tag{46}$$

The velocities after collision can be obtained by using Eqs. (30) and (31). The diameters of droplets after collision are unaltered.

Case C_{VS} > 0: the separation volumes from droplets determine the evolution of the temporary fluid ligament that would form between them. In this model, it is assumed that the ligament has a uniform cylindrical shape, and the radius r_o of ligament at initial instant time of the temporal evolution with a momentum balance equation can be obtained:

$$\frac{1}{6}\pi\left(\Psi_S\phi_S^3 + \Psi_L\phi_L^3\right) = \pi r_0^2\eta, \tag{47}$$

where Ψ_S and Ψ_L are the fraction of volumes lost from the smaller and large droplets to form the ligament [17, 18], and η is its initial time instantaneous length (**Figure** 7). Another assumption is $\eta = r_o$. In this model, a time scale of temporal evolution ligament is proposed: $T = 0.75k_2\sqrt{\mathrm{We}_0}$. If $T \le 2$, the ligament contracts in a single satellite whose radius is r_o. $k_2 = 0.45$ and $\mathrm{We}_0 = 2r_0(\rho_l/\sigma)\left|\vec{V}_R\right|^2$. Otherwise, it is necessary to compute the evolution time of ligament radius equation for obtaining the final value r_{bu}:

$$\frac{3}{4\sqrt{2}}k_1k_2\sqrt{\mathrm{We}_0}\left(\frac{r_{bu}}{r_0}\right)^{7/2} + \left(\frac{r_{bu}}{r_0}\right)^2 - 1 = 0, \tag{48}$$

$k_1 = 11.5$ following Kim [18]. Eq. (48) can be solved by iteration with an initial value $(r_{bu}/r_0) = 1$ and $\Delta t = 1 \times 10^{-2}$.

The diameter of satellite droplets can be determined by following Georjon [41]:

$$\phi_{sat} = 3.78r_{bu}. \tag{49}$$

The number of satellite droplets is calculated from the mass conservation by assuming uniform satellites size $N_{sat} = 6(r_0/\phi_{sat})^3$. The velocities of satellite droplets can be obtained from momentum equation where the velocities of parent droplets after collision are computed by Eq. (46).

$$\vec{V}_{sat} = \frac{\left(\phi_L^3\vec{V}_L - \phi_{L\ (new)}^3\vec{V}_{L\ (new)}\right) + \left(\phi_S^3\vec{V}_S - \phi_{S\ (new)}^3\vec{V}_{S\ (new)}\right)}{N_{sat}\phi_{sat}^3}, \tag{50}$$

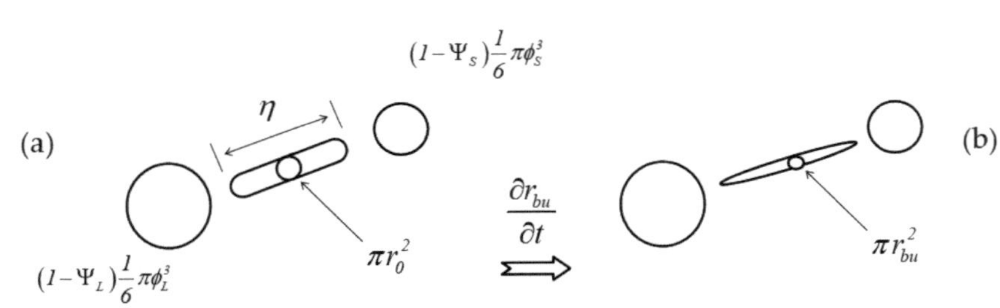

Figure 7.
Collision model for the stretching outcome. (a) Formation instant time of ligament and (b) temporal evolution ligament.

where the diameters of parent droplets after collision are:

$$\begin{cases} \phi_{L\ (new)} = (1 - \Psi_L)\dfrac{1}{6}\pi\phi_L^3 \\ \phi_{S\ (new)} = (1 - \Psi_S)\dfrac{1}{6}\pi\phi_S^3 \end{cases} . \tag{51}$$

2.3.4.3 Reflexive

The volume of ligament is the entire temporarily merged mass of two droplets. The model of satellite droplet formation is similar at stretching outcome, but the initial radius of ligaments is $r_0^3 = (\phi_L/2)^3 + (\phi_S/2)^3$. The model proposed by Munnannur [17] for reflexive outcome uses the time scale of temporal evolution ligament. When $T \leq 3$, a single satellite droplet is formed and the three droplets (considering the ligament breaks up into two end-droplets and one-satellite droplet) have same size. However, according to the experimental studies of Ashrgiz [34], Kim [18] affirms that no uniform droplet sizes are obtained for the end-droplets and a single satellite droplet after reflexive collision. We have adopted the last criteria, so the satellite droplet diameters are computed with Eqs. (48) and (49), but the number of satellite droplets is $N_{sat} = 6(r_0/\phi_{sat})^3 - 2$. If $N_{sat} \leq 0$, it is assumed that the ligament breaks up without satellite droplet and the two end-droplets have their own radius. If $0 < N_{sat} \leq 1$, it is assumed that a single satellite droplet is formed that is smaller than the two end-droplets after collision. The diameter of single satellite droplets is ϕ_{sat} and the end-droplets after collision have identical diameters:

$$\phi_{L\ (new)} = \phi_{S\ (new)} = \left(\frac{8r_0^3 - \phi_{sat}^3}{2}\right)^{1/3} . \tag{52}$$

When $N_{sat} > 1$, the ligament breaks up into uniform droplets with identical diameters $\phi_{L\ (new)} = \phi_{S\ (new)} = \phi_{sat}$. The velocities of end-droplets and satellite droplets are computed with Eqs. (46) and (50).

2.3.4.4 Bounce

In this case, the droplets bounce maintaining their diameters after the impact. In the general case, oblique collision between droplets is considered. The droplet velocities after collision must be decomposed into a normal component and a tangential component to the plane of impact. The tangential component after impact remains unchanged, but the normal component is affected by a soft inelastic rebound assuming a restitution coefficient $e_{n,p} = 0.97$ by following Almohammed [42]. This restitution coefficient takes into account the dissipation of kinetic energy during the impact. The normal velocities of droplets at instant of collision are as follows:

$$\begin{cases} V_{n\ L} = V_{i\ L}\dfrac{|x_{i\ R}|}{b} \\ V_{n\ S} = V_{i\ S}\dfrac{|x_{i\ R}|}{b} \end{cases} , \tag{53}$$

where $x_{i\ R} = x_{i\ S} - x_{i\ L}$ is the relative position between the droplets in $i = 1, 2, 3$ (x, y, z) and b is the center droplet distance (**Figure 2**). The normal component velocities after collision are given as:

$$\begin{cases} V_{n\,L\,(new)} = V_{n\,L} - \dfrac{\Delta^3}{1+\Delta^3} f_n^{dd} \\ V_{n\,S\,(new)} = V_{n\,S} + \dfrac{1}{1+\Delta^3} f_n^{dd} \end{cases}, \tag{54}$$

where f_n^{dd} is the normal impulse of a droplet-droplet collision: $f_n^{dd} = -(1+e_{n,p})(V_{n\,S} - V_{n\,L})$.

3. Results and discussion

3.1 Ejection of liquid particle simulation

In order to obtain the Weibull p.d.f. corresponding to the droplet diameters as described in Section 2.1, the scale k = 301.228 and shape m = 2.606 parameters were obtained with R-R method. **Figure 8** shows the minimum square adjusted of regression line with a correlation coefficient R^2 = 0.9965.

3.2 Effects of the Eulerian-Lagrangian double-way coupling

The trajectories of liquid particles are simulated with an Euler-Lagrangian double-way coupled model descript in Section 2.2. The influence of droplets to air velocity is shown in **Figure 9** (a) for t = 1 s instant time and (b) for t = 20 s instant time of the simulation. The vertical velocities of air W are shown in color scale. Eddies around the spray plume are formed and extend up to 3 m from the center of the spray (c). It is observed that eddies are formed by influence of jet droplets. This effect should be taken into account as droplets of very small diameters are captured by eddies and are prone to contribute to drift. This effect can be seen in **Figure 10** where the small droplets follow the streamlines of eddies on both sides of the sprayer.

The results of the vertical droplet velocities distribution as a function of the droplet diameters obtained at 0.35 m below the nozzle are shown in **Figure 11**. These are compared with the laboratory measurements of Nuyttens [12]. It is

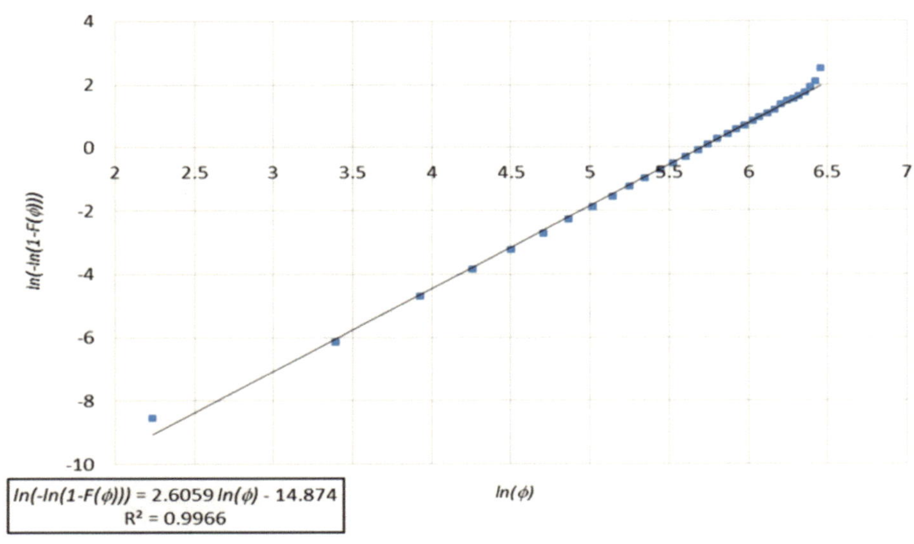

Figure 8.
Regression line of R-R distribution function and data measurement of droplet diameters.

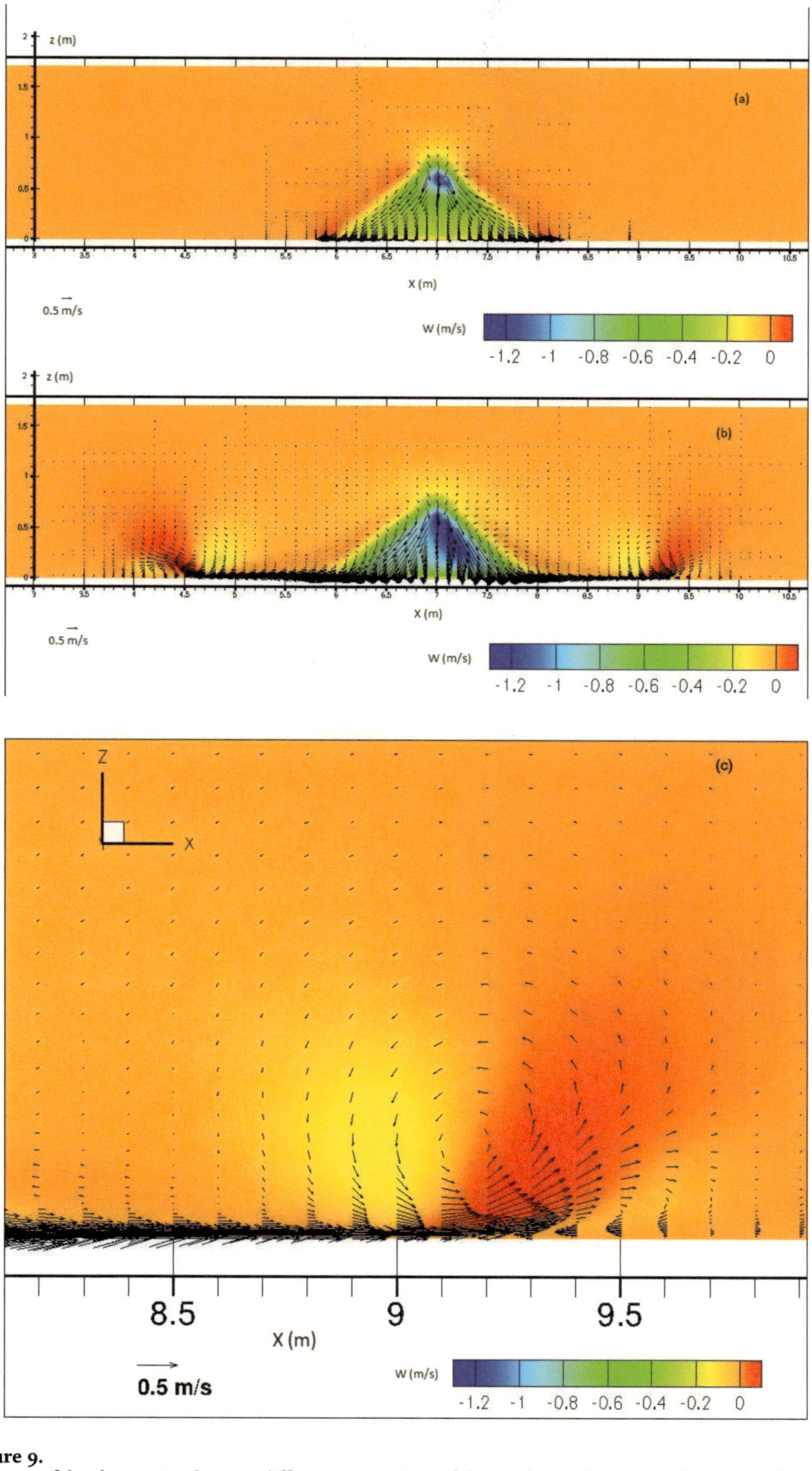

Figure 9.
Influence of droplets to air velocity at different instant times of the simulation. (a) t = 1 *s, (b)* t = 20 *s, (c) zoom of eddy formed at 2.5 m from center spray at 20 s, and* W *is a vertical component of air velocity around the spray.*

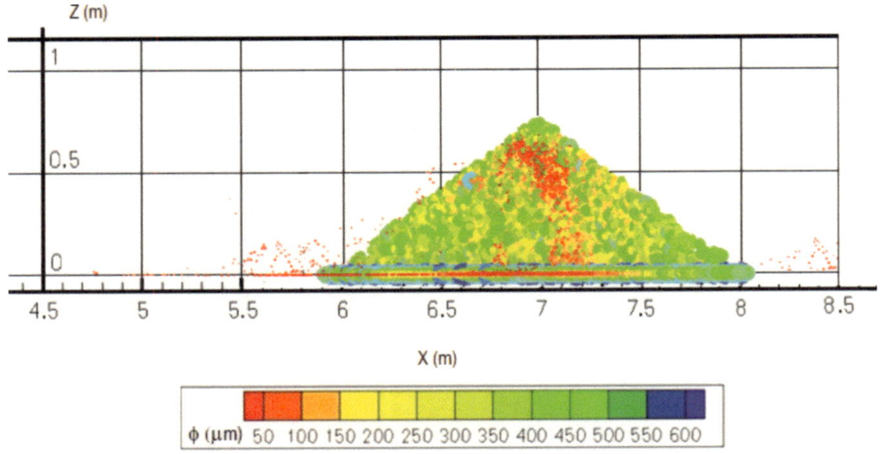

Figure 10.
Position of droplets at t = 20 s *of the simulation classified by their diameters.*

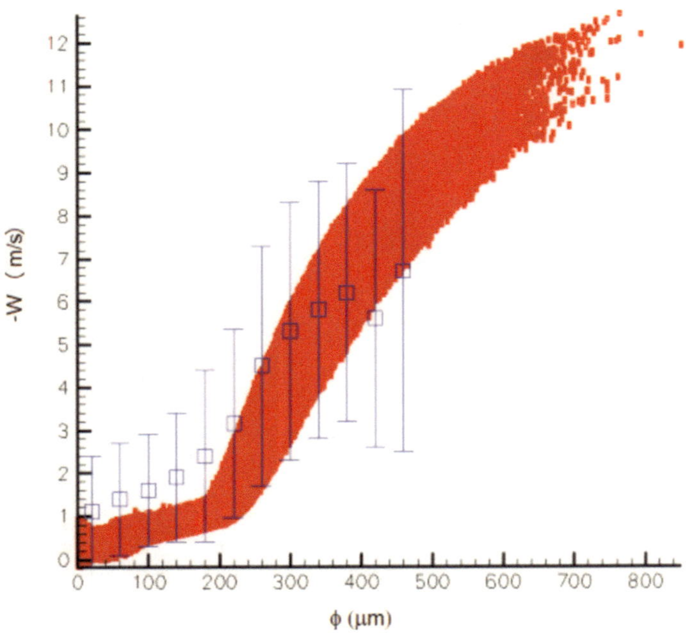

Figure 11.
Distribution of vertical droplet velocities in (m/s) as a function of the diameters (μm). ■ *Droplet simulation.* ├─□─┤ *Mean and extreme range values measured by Nuyttens [12].*

observed that the dispersion of velocity values for each diameter class is greater in laboratory measurements than in simulation. In addition, for diameters less than 200 μm, the model slightly underestimates the vertical velocity values relative to the laboratory results.

3.3 Binary collision droplet map

The collision map for binary droplet model descripted in Section 2.3 is shown in **Figure 12**. The map allows showing the events of coalescence, bounce, reflexive,

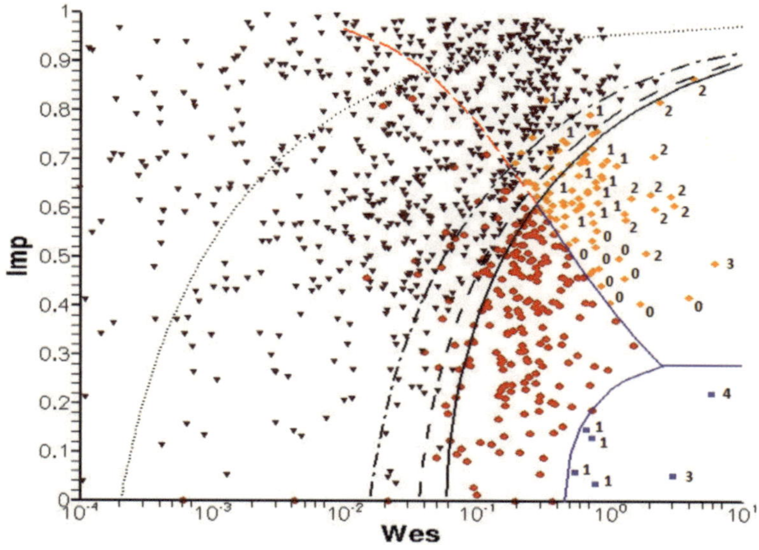

Figure 12.
Map outcomes from binary droplet collision model. ● Coalescence, ■ reflexive, ◆ stretching, and ▼ bounce. The numbers next to the symbols indicate the number of satellite droplets formed.

and stretching separation. When considering the total number of droplet binary collision events, 21.1% corresponds to coalescence, 0.6% to reflexive separation, 8.8% to stretching separation, and 69.5% to bounce. The amount of satellite droplets arising from the separation by reflexive and stretching is displayed with numbers. It is noted that the number of satellite drops increases with the number of symmetrical Weber for both separately. This behavior indicates that the greater velocity the droplets are ejected from the spray nozzle, the more likely it is that satellite droplets will appear as a result of reflexive and stretching separation. As mentioned above,

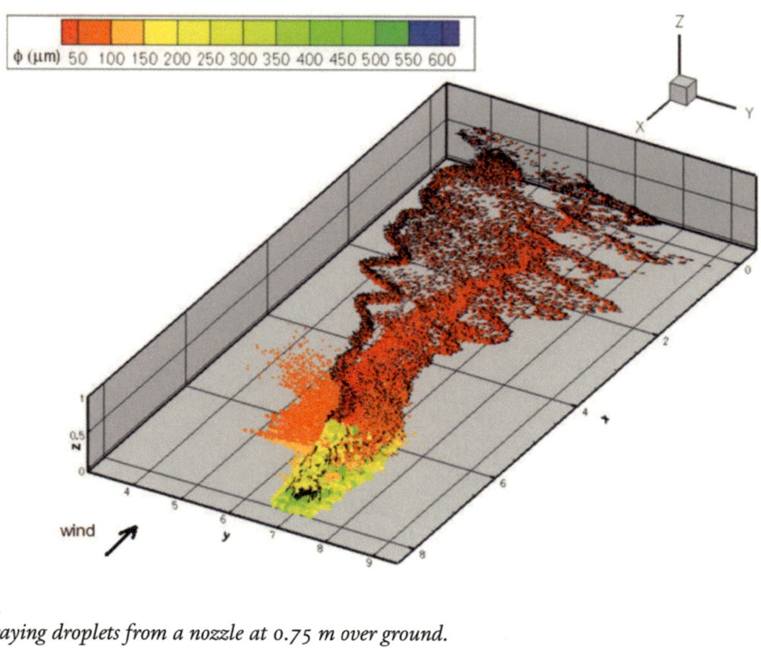

Figure 13.
Drift of spraying droplets from a nozzle at 0.75 m over ground.

this can cause an increase in the proportion of sprayed product not reaching its destination, leaving it adrift.

3.4 Simulation of droplet dispersion from a nozzle in a cultivated field

Figure 13 shows the drift simulation of droplets spraying over cultivate field with a nozzle at 0.75 m above the ground.

The meteorological conditions of air temperature at nozzle level are 30°C with 2 m s^{-1} velocity wind. The simulation time shown in **Figure 13** is 20 s after the start of spraying. The drift of small droplets (less than 50 μm in diameter) exceeds 8 m in the area of application. Of the total liquid sprayed, 0.43% corresponds to droplets smaller than 50 μm measured by Nuyttens [12] in wind tunnel at 50 cm below the spray nozzle. In the simulation shown in **Figure 13**, this percentage does not change because the satellite droplets generated are greater than 80 μm. The number of satellite droplets generated by stretching and reflective separation in these conditions was obtained. Of the 120 satellite droplets analyzed, 35.3% have diameters less than 150 μm, 61.3% have diameters between 150 and 250 μm, and 3.4% have diameters between 250 and 350 μm. There were no satellite droplets with diameters larger than 350 μm.

4. Conclusions

In the present work, it was possible to simulate and validate the ejection velocity of the liquid particles from an HARDI™ ISO F110 03 nozzle placed at 0.75 m over ground. The diameters of the drops were randomized to the volume applied following a procedure of Rosin-Rammler distribution function for obtaining the parameters of Weibull probability density function with a correlation coefficient $R^2 = 0.997$. The double-way coupled Euler-Lagrangian model has been used for obtaining the trajectory of droplet spraying. Eddies at both sides of spraying have been captured by the model. These extend up to 3 m from the center of the spray. The vertical component droplet velocity was simulated and validated with laboratory measurements. The velocity of droplets smaller than 200 **μ**m slightly underestimates with respect to laboratory data. The binary collision models have been implemented into the code to consider particle collision events. A collision detection algorithm using collision boxes was presented and used to optimize computation times. The drift of droplets with air temperature of 30°C and wind speed value 2 m s^{-1} has been simulated in cultivated fields. The drift of droplets smaller than 50 **μ**m diameter exceeds 8 m of the application area. No satellite droplets smaller than 80 **μ**m are generated under field simulation conditions. The largest proportion of satellite droplets generated as a result of the droplet collision has a diameter between 80 and 250 **μ**m.

Acknowledgements

The present work is funded through the PIO CONICET-UNER 2015–2016 project and the FCyT-UADER research project “Development of a simulation model for the study of the drift of agricultural sprayings, using a flat fan nozzle, from trailing equipment." We are also grateful to the Laboratory of Prototyping of Electronics and 3D Printing of the School of Engineering, UNER, for the work of printing the nozzle.

A. Appendix A

The Weibull probability density function (Weibull p.d.f.) can be used for describing a lot of technical applications for which the distribution of ground material, particles dispersion, or droplet diameters in spray jet normally in the μm-band have behaviors with a random characteristic. In this case, the diameters of droplets ejected from a spray nozzle are simulated using a Weibull p.d.f.

Let us name the random variable ϕ_0 that represents the initial diameter of the liquid particles, and the expression of Weibull p.d.f. for this random variable is:

$$\begin{cases} f(\phi_0) = \frac{m}{k^m}\phi_0^{m-1}\exp\left[-\left(\frac{\phi_0}{k}\right)^m\right], & \text{if } \phi_0 \geq 0, \\ f(\phi_0) = 0, & \text{if } \phi_0 < 0. \end{cases} \tag{55}$$

In this p.d.f., $m > 0$ is the shape parameter and $k > 0$ is the scale parameter. The form of the density function of the Weibull distribution changes drastically with the value of m parameters. The k parameter does not change the shape of the distribution, but it extends along the random variable ϕ_0. In this way, if the parameters m and k are chosen correctly, it is possible to obtain the shape and stretch of the Weibull p.d.f. that fits the experimentally measured diameter data.

To take these data into account, the cumulative function of the Weibull distribution, named Rosin-Rammler (R-R), is used.

The R-R distribution function $F(\phi_0)$ is expressed as:

$$F(\phi_0) = 1 - \exp\left[-\left(\frac{\phi_0}{k}\right)^m\right]. \tag{56}$$

Eq. (A.2) can be written as:

$$\text{Ln}\{-\text{Ln}[1 - F(\phi_0)]\} = m.\text{Ln}(\phi_0) - m.\text{Ln}(k). \tag{57}$$

If we associate Eq. (A.3) with a linear equation $f(X) = mX + a$ where the dependent variable is $f(X) = \text{Ln}\{-\text{Ln}[1 - F(\phi_0)]\}$ and the independent variable is $X = \text{Ln}(\phi_0)$, m is the slope and $a = -m\text{Ln}(k)$ is intercept, in which the line cuts on the $Y = f(X)$ axis. The method to obtain the Weibull p.d.f. parameters consists of plotting over logarithmic scale; the results of the experimental measurements $\text{Ln}\{-\text{Ln}[1 - F(\phi_0)]\}$ vs. $\text{Ln}(\phi_0)$and approximate the point cloud to a linear regression by the least squares to obtain the slope (shape parameter) m and the constant term a. The k scale parameter is obtained by $k = \exp(-a/m)$.

With these two parameters (m, k), it is possible to obtain the Weibull p.d.f. that describes the droplet diameter's distribution corresponding to the experiment measurements.

Author details

Carlos G. Sedano[1], César Augusto Aguirre[2,3]* and Armando B. Brizuela[1,2,3]

1 School of Science and Technology, Autonomous University of Entre Ríos (FCyT-UADER), Entre Ríos, Argentina

2 School of Agricultural Sciences, National University of Entre Ríos (FCA-UNER), Entre Ríos, Argentina

3 National Research Council of Science and Technology (CONICET), Buenos Aires, Argentina

*Address all correspondence to: cesaraguirredalotto@gmail.com

References

[1] Battistoni M, Xue Q, Som S. Large-eddy simulation (LES) of spray transients: Start and end of injection phenomena. Oil & Gas Science and Technology—Rev IFP Energies Nouvelles. 2015;**2015**:1-24. DOI: 10.2516/ogst/2015024

[2] Sabelnikov V, Lipatnikov A, Chakraborty N, Nishiki S, Hasegawa T. A transport equation for reaction rate in turbulent flow. Physics of Fluids. 2016; **28**(8):081701. DOI: 10.1063/1.4960390

[3] Wei G, Vinkovic I, Shao L, Simoëns S. Scalar dispersion by a large-eddy simulation and a Lagrangian stochastic subgrid model. Physics of Fluids. 2006; **18**:095101. DOI: 10.1063/1.2337329

[4] Aguirre C, Brizuela A, Vinkovic I, Simoëns S. A subgrid Lagrangian stochastic model for turbulent passive and reactive scalar dispersion. International Journal of Heat and Fluid Flow. 2006;**27**(4):627-635. DOI: 10.1016/j.ijheatfluidflow.2006.02.011

[5] Vinkovic I, Aguirre C, Simoëns S. Large-eddy simulation and Lagrangian stochastic modelling of passive scalar dispersion in a turbulent boundary layers. Journal of Turbulence. 2006; 7(30):1468-5248

[6] Vinkovic I, Aguirre C, Simoëns S, Gorokhovski M. Large-eddy simulation of droplet dispersion for inhomogeneous turbulent wall flow. International Journal of Multiphase Flow. 2006;**32**(3):344-364

[7] Vinkovic I, Aguirre C, Ayrault M, Simoëns S. Large-eddy simulation of the dispersion of solid particles in a turbulent boundary layers. Journal of Boundary Layers-Meteorology. 2006:1472-1573. DOI: 10.1007/s10546-006-9072-6

[8] Stephens DW, Keough S, Sideroff C. Euler-Lagrange large eddy simulation of a square cross-sectioned bubble column. In: APCChE 2015 Congress incorporating Chemeca; 27 Sept–01 Oct 2015; Melbourne, Victoria; Paper no. 3134665. 2015. pp. 1-12

[9] Zamansky R, Vinkovic I, Gorokhovsky M. Acceleration in turbulent channel flow: Universalities in statistic, subgrid stochastic models and an application. Journal of Fluid Mechanics. 2013;**721**:627-668. DOI: 10.1017/jfm.2013.48

[10] Orcellet E, Berri J, Aguirre C, Müller G. Atmospheric dispersion study of TRS compounds emitted from a pulp mill plant in coastal regions of the Uruguay river, South America. Aerosol and Air Quality Research. 2016;**16**(6): 1473-1482. DOI: 10.4209/aaqr.2015.02.0112

[11] Aguirre C, Brizuela A. Computational tools for the simulation of atmospheric pollution transport during a severe wind event in Argentina. In: Atmospheric Hazards—Case Studies in Modeling, Communication, and Societal Impacts. Chapter 6. InTech Open Science, Open Mind.; 2016. pp. 111-136. DOI: 10.5772/63552

[12] Nuyttens D. Drift from field crops prayers: The influence of spray application technology determined using indirect and direct drift assessment means [thesis]. Germany: Faculteit Bio-ingenieurs wetenschappen, Katholieke Universiteit Leuven; 2007

[13] Xue M, Droegemeier K, Wong V. The advanced regional prediction system (ARPS). A multi-scale nonhydrostatic atmospheric simulation and prediction model. Part I: Model dynamics and verification. Meteorology Atmospheric Physics. 2000;**75**:161-193

[14] Aguirre C. Dispersión et Mélange Atmosphérique Eulero-Lagrangien de Particules Fluides Réactives. Application à des cas simples et complexes [thesis]. Lyon, France: Université Claude Bernard; 2005. 337 p

[15] Fackrell J, Robins A. Concentration fluctuation and fluxes in plumes from point sources in a turbulent boundary layers. Journal of Fluid Mechanics. 1982; **117**:1-26

[16] Gong W. A wind tunnel study of turbulent dispersion over two- and three-dimensional gentle Hills from upwind point sources in neutral flow. Boundary Layers Meteorology. 1991;**54**: 211-230

[17] Munnannur A, Reitz R. A new predictive model for fragmenting and non-fragmenting binary droplet collisions. International Journal of Multiphase Flow. 2007;**33**(8):873-896. DOI: 10.1016/j.ijmultiphaseflow.2007.03.003

[18] Kim S, Lee D, Lee C. Modeling of binary droplet collisions for application to inter-impingement sprays. International Journal of Multiphase Flow. 2009;**35**:533-549. DOI: 10.1016/j.ijmultiphaseflow.2009.02.010

[19] Zhang H, Li Y, Li J, Liu Q. Study on separation abilities of moisture separators based on droplet collision models. Nuclear Engineering and Design. 2017;**325**:135-148. DOI: 10.1016/j.nucengdes.2017.09.030

[20] Rabe C, Malet J, Feuillebois F. Experimental investigation of water droplet binary collisions and description of outcomes with a symmetric Weber number. Physics of Fluids. 2010;**22**: 047101. DOI: 10.1063/1.3392768

[21] Sedano C, Aguirre C, Brizuela A. Simulación de la Eyección de Spray Líquido desde un Pico de Pulverizadora para Aplicación de Herbicidas. Asociación Argentina de Mecánica Computacional AMCA - Mecánica Computacional. 2017;**VXXIII**: 1049-1068

[22] Ayres D, Caldas M, Semião V, da Graça Carvalho M. Prediction of the droplet size and velocity joint distribution for sprays. Fuel. 2001;**80**: 383-394

[23] Baetens K. Development and application of drift prediction models in fields praying [thesis]. Leuven: Faculteit Bio-ingenieurs wetenschappen, Katholieke Universiteit; 2009

[24] Macías-García A, Cuerda-Correa E, Díaz-Díez M. Application of the Rossin-Rammler and Gates-Gaudin-Schuhmann models to the particle size analysis of agglomerated cork. Material Characterization. 2004;**52**:159-164. DOI: 10.1016/j.matchar.2004.04.007

[25] Michelot C. Développement d'un modèle stochastique lagrangien. Application à la dispersion et à la chimie de l'atmosphère [thesis]. Lyon, France: Université Claude Bernard; 1996. 180 p

[26] Holterman H. Kinetic and evaporation of waterdrops in air, Wageningen: IMAG. Report 2003-12; Wageningen UR, InstituutvoorvMilieu en Agritechniek; 2003. 67 p

[27] Sedano C, Aguirre A, Brizuela B. Numerical simulation of spray ejection from nozzle for herbicide application: Comparison of drag coefficient expressions. Computers and Electronics in Agriculture. Elsevier. Forthcoming

[28] Turton R, Levenspiel O. A short note on the drag correlation for sphere. Powder Technology. 1986;**47**:83-86. DOI: 10.1016/0032-5910(86)80012-2

[29] Germano M, Piomelli U, Moin P, Cabot WH. A dynamic subgrid-scale eddy viscosity model. Physics of Fluids. 1991;**A**(3):1760-1765

[30] Gicquel LYM, Givi P, Jaberi FA, Pope SB. Velocity filtered density function for large-eddy simulation of turbulent flow. Physics of Fluids. 2002; **14**(3):1196-1213

[31] Deardorff JW. Stratocumulus-capped mixed layer derived from a three dimensional model. Boundary Layer Meteorology. 1980;**18**:495-527

[32] Pope SB. Lagrangian PDF methods for turbulent flows. Annual Review of Fluid Mechanics. 1994;**26**:23-63

[33] Brazier-Smith PR, Jennings SG, Latham J. The interaction of falling water drops: Coalescence. Proceedings of the Royal Society of London. Series A, Mathematical and Physical Sciences. 1972;**326**(1566):393-408. DOI: 10.1098/rspa.1972.0016

[34] Ashrgiz N, Poo JY. Coalescence and separation in binary collisions of liquid drops. Journal of Fluid Mechanics. 1990; **221**:183-204

[35] Estrade JP, Carentz H, Lavergne G, Biscos Y. Experimental investigation of dynamic binary collision of ethanol droplets—A model for droplet coalescence and bouncing. International Journal of Heat and Fluid Flow. 1999; **20**(5):486-491. DOI: 10.1016/S0142-727X(99)00036-3

[36] Brenn G, Valkovska D, Danov K. The formation of satellite droplets by unstable binary drop collisions. Physics of Fluids. 1994;**13**(9):2463, 2001-2477. DOI: 10.1063/1.1384892

[37] Kollár LE, Farzaneh M, Karev AR. Modeling droplet collision and coalescence in an icing wind tunnel and the influence of these processes on droplet size distribution. International Journal of Multiphase Flow. 2005;**31**(1): 69-92. DOI: 10.1016/j.ijmultiphaseflow.2004.08.007. ISSN 0301-9322

[38] Ko GH, Ryou HS. Modeling of droplet collision-induced breakup process. International Journal of Multiphase Flow. 2005;**31**:723-738. DOI: 10.1016/j.ijmultiphaseflow.2005.02.004. ISSN 0301-9322

[39] Hu C, Xia S, Li C, Wu G. Three-dimensional numerical investigation and modeling of binary alumina droplet collisions. International Journal of Heat and Mass Transfer. 2017;**113**:569-588. DOI: 10.1016/j.ijheatmasstransfer.2017.05.094

[40] O'Rourke P. Collective drop effects in vaporizing liquid sprays [PhD dissertation]. Princeton, NJ: Dept. Mech. Aerospace Engg., Princeton University; 1981

[41] Georjon TL, Reitz RD. A drop-shattering collision model for multidimensional spray computations. Atomization and Sprays. 1999;**9**(3)

[42] Almohammed N. Modelling and simulation of particle agglomeration droplet coalescence and particle-wall adhesion in turbulent multiphase flow [thesis]. Hamburg, Germany: Helmet-Schmidt University; 2018; 402 p

Section 2

Computational Fluid Dynamics Applications

Chapter 4

Scalar Conservation Laws

Baver Okutmuştur

Abstract

We present a theoretical aspect of conservation laws by using simplest scalar models with essential properties. We start by rewriting the general scalar conservation law as a quasilinear partial differential equation and solve it by method of characteristics. Here we come across with the notion of strong and weak solutions depending on the initial value of the problem. Taking into account a special initial data for the left and right side of a discontinuity point, we get the related Riemann problem. An illustration of this problem is provided by some examples. In the remaining part of the chapter, we extend this analysis to the gas dynamics given in the Euler system of equations in one dimension. The transformations of this system into the Lagrangian coordinates follow by applying a suitable change of coordinates which is one of the main issues of this section. We next introduce a first-order Godunov finite volume scheme for scalar conservation laws which leads us to write Godunov schemes in both Eulerian and Lagrangian coordinates in one dimension where, in particular, the Lagrangian scheme is reformulated as a finite volume method. Finally, we end up the chapter by providing a comparison of Eulerian and Lagrangian approaches.

Keywords: conservation laws, Burgers' equation, shock and rarefaction waves, weak and strong solutions, Riemann problem, Euler system, Godunov schemes, Eulerian coordinates, Lagrangian coordinates

1. Introduction

We present a general form of scalar conservation laws with further properties including some basic models and provide examples of computational methods for them. The equations described by

$$\partial_t u + \partial_x f(u) = 0, \quad t > 0, x \in \mathbb{R} \tag{1}$$

in one dimension are known as scalar conservation laws where $u = u(t, x)$ is the conserved quantity and $f = f(u)$ is the associated flux function depending on t and x. Whenever an initial condition $u(0, x) = u_0(x)$ is attached to Eq. (1), the problem is called the Cauchy problem the solution of which is a content of this chapter. The outlook of chapter is as follows. We introduce basic concepts and provide particular examples of scalar conservation laws in the first part. The equation of gas dynamics in Eulerian coordinates in one dimension is the main issue

IntechOpen

of the second part. After providing further instruction for these equations, we provide a transformation of the Eulerian equations in the Lagrangian coordinates. In the final part, we give as an example of computational methods for conservation laws, the Godunov schemes for the Eulerian, and the Lagrangian coordinates, respectively.

1.1 Conservation laws: integral form and differential form

We start by investigating the relation of the equations in gas dynamics with conservation laws. We take into account the equation of conservation of mass in one dimension. The density and the velocity are assumed to be constant in the tube where x is the distance and $\rho(t,x)$ is the density at the time t and at the point x. Then if we integrate the density on $[x_1,x_2]$, we get total mass $\int_{x_1}^{x_2} \rho(t,x)dx$ at time t. Assigning the velocity by $u(t,x)$, then mass flux at becomes $\rho(t,x)u(t,x)$. It follows that the rate of change of the mass in $[x_1,x_2]$ is

$$\frac{d}{dt}\int_{x_1}^{x_2} \rho(t,x)dx = \rho(t,x_1)u(t,x_1) - \rho(t,x_2)u(t,x_2). \tag{2}$$

The last equation is called integral form of conservation law. Integrating this expression in time from t_1 to t_2, we get

$$\int_{x_1}^{x_2} \rho(t_2,x)dx - \int_{x_1}^{x_2} \rho(t_1,x)dx = \int_{t_1}^{t_2} \rho(t,x_1)u(t,x_1)dt - \int_{t_1}^{t_2} \rho(t,x_2)u(t,x_2)dt. \tag{3}$$

Using the fundamental theorem of calculus after reduction of Eq. (3), it follows that

$$\rho(t,x_2)u(t,x_2) - \rho(t,x_1)u(t,x_1) = \int_{x_1}^{x_2} \partial_x(\rho(t,x)u(t,x))dx. \tag{4}$$

As a result, we get

$$\int_{t_1}^{t_2}\int_{x_1}^{x_2} \{\partial_t\rho(t,x) + \partial_x\rho(t,x)u(t,x)\}\, dx\, dt = 0. \tag{5}$$

Here the end points of the integrations are arbitrary; that is, for any $[x_1,x_2]$ and $[t_1,t_2]$, the integrant must be zero. It follows that the conservation of mass yields

$$\partial_t\rho + \partial_x(u\rho) = 0, \tag{6}$$

which is said to be the differential form of the conservation law.

1.2 A first-order quasilinear partial differential equations

A general solution to a quasilinear partial differential equation of the form

$$a(t,x,u)\partial_t u + b(t,x,u)\partial_x u = c(t,x,u) \tag{7}$$

where a, b, c are non-zero and smooth on a given domain $D \in \mathbb{R}^3$ follows by the characteristic method where the characteristic curves are defined by

$$\frac{dt}{a(t,x,u)} = \frac{dx}{b(t,x,u)} = \frac{du}{c(t,x,u)}. \tag{8}$$

By applying a parametrization of c, the relation (8) is transformed to a system of ordinary differential equation (ODE):

$$\frac{dt}{dc} = a(t,x,u), \quad \frac{dx}{dc} = b(t,x,u), \quad \frac{du}{dc} = c(t,x,u). \tag{9}$$

In addition to these equations, if an initial condition $u_0 = u(x_0)$ is also given, then by the existence theorem of ODE, there is a unique characteristic curve passing from each point (t_0, x_0, u_0) leading to an integral surface which is the solution to Eq. (7).

Observe that the scalar conservation law (1) is a particular example of Eq. (7) if we assign $a(t,x,u) = 1$, $b(t,x,u)u_x = (f(u))_x$, and $c(t,x,u) = 0$. The conserved quantity can be observed by integrating equation (1) over $[x_0, x_1]$. Indeed

$$\begin{aligned} \frac{d}{dt}\int_{x_0}^{x_1} u(t,x)dx &= \int_{x_0}^{x_1} \partial_t u(t,x)dx = -\int_{x_0}^{x_1} f(u(t,x))_x dx \\ &= f(u(t,x_1)) - f(u(t,x_0)) \\ &= [\text{inflow at the point} \, x_1] - [\text{outflow at the point} \, x_0]. \end{aligned} \tag{10}$$

This means, the quantity $u(t,x)$ is conserved so that it depends on the difference of the flux functions between the points x_0 and x_1.

1.3 Strong (classical) solutions

We consider the initial value problem

$$\begin{aligned} &\partial_t u + \partial_x(f(u)) = 0, \quad t > 0, \quad x \in \mathbb{R} \\ &u(0,x) = u_0(x), \quad x \in \mathbb{R} \end{aligned} \tag{11}$$

where the initial data is assumed to be continuously differentiable, that is, $u_0(x) \in C^1(\mathbb{R})$. Applying the chain rule to the relation (11), it follows that

$$\begin{aligned} &\partial_t u + f'(u)\partial_x u = 0, \quad t > 0, \quad x \in \mathbb{R}, \\ &u(0,x) = u_0(x), \quad x \in \mathbb{R}, \end{aligned} \tag{12}$$

where we define characteristic curves of Eq. (12) to be the solution of $\frac{d}{dt}x(t) = f'(u(t,x(t))) = f'(u)$. Then a solution to the system (12) in a domain $\Omega \in \mathbb{R}$ is said to be a strong (or classical) solution if it satisfies Eq. (11), and it is continuously differentiable on a domain $\Omega \in \mathbb{R}$. Let u be a strong solution and the initial data u_0 be differentiable. Observe that (12) is equivalent to a quasilinear form:

$$\partial_t u + \lambda(u)\partial_x u = 0, \tag{13}$$

with $\lambda(u) = f'(u)$. Applying the method of characteristics to Eq. (13), the partial differential equation is transformed to a system of ordinary differential equations. We consider the characteristic curve passing through the point $(0, x_0)$:

$$\begin{aligned} \partial_t x &= \lambda(u(t, x(t))) \\ x(0) &= x_0. \end{aligned} \tag{14}$$

Along this characteristic curve,

$$\partial_t u(t, x(t)) = \partial_t u(t, x(t)) + \partial_t x \partial_x u(t, x(t)) = \partial_t u + \lambda(u)\partial_x u = 0 \tag{15}$$

is satisfied, that is, u is constant. Hence, the characteristic curves are straight lines satisfying

$$x = x_0 + \lambda(u_0(x_0))t = 0. \tag{16}$$

Hence we can define smooth solutions by $u(t, x) = u_0(x_0)$. If the slope of the characteristics is $m_{char} = \frac{1}{\lambda(u_0(x_i))}$, then depending on the behavior of λ, the solution takes different forms. If $\lambda(u_0(x))$ is increasing, then the slopes of the characteristics are decreasing. As a result, the characteristics do not intersect, and thus solution can be defined for all t which is greater than zero. On the other hand, if $\lambda(u_0(x))$ is decreasing, then the slopes of the characteristics will be increasing which implies that the characteristics intersect at some point. But at the intersection point, solution cannot take both values $u_0(x_1)$ and $u_0(x_2)$. Therefore, we cannot define the strong solution for all $t > 0$.

1.4 Linear advection equation

The basic example of the scalar conservation law is the linear advection equation. It can be obtained by setting $a(t, x, u) = 1$, $b(t, x, u) = \lambda$, and $c(t, x, u) = 0$ in Eq. (7). The flux function takes the form $f(u) = \lambda u$ where λ is a constant. Then the following quasilinear partial differential equation

$$\partial_t u + \lambda \partial_x u = 0 \tag{17}$$

is a linear advection equation. Similar to Eqs. (11) and (12), an initial value problem for linear advection equation is described by

$$\begin{aligned} &\partial_t u + \partial_x f(u) = 0, \quad -\infty < x < \infty, \quad t \geq 0, \\ &u(0, x) = u_0(x) = f(x_0), \quad -\infty < x < \infty. \end{aligned} \tag{18}$$

Applying the method of characteristics, it follows that $\frac{dt}{1} = \frac{dx}{\lambda} = \frac{du}{0}$ or equivalently

$$u = c_1, \quad \frac{dx}{dt} = \lambda = c_1, \quad x = c_1 t + c_2, \tag{19}$$

where c_1 and c_2 are constant and $x - \lambda t = c_2$. As a conclusion, the solution is

$$u(t, x) = u_0(x - \lambda t), \quad t \geq 0. \tag{20}$$

Here λ is the wave speed, and the characteristic lines $x - \lambda t = c_2$ are wavefronts which are constants.

1.5 Burgers' equation

Burgers' equation is the simplest nonlinear partial differential equation and is the one of the most common models used in the scalar conservation laws and fluid dynamics. The classical Burgers' equation is described by

$$\partial_t u + u\partial_x u = \nu \partial_{xx} u, \tag{21}$$

where $\nu\partial_{xx}u$ is the viscosity term. Equation (21) can be considered as a combination of nonlinear wave motion and linear diffusion term so that it is balance between time evolution, nonlinearity, and diffusion. The term $u\partial_x u$ is a convection term that may have an effect to wave breaking, and the term $\nu\partial_{xx}u$ is a diffusion term that may cause to efface the wave breaking and to flatten discontinuities, and thus we expect to achieve a smooth solution. We try to find a traveling wave solution of Eq. (21) of the form

$$u(t,x) = g(\xi) = g(x - \lambda t), \quad \text{with} \quad \xi = x - \lambda t, \tag{22}$$

where g and λ are to be determined. Applying the chain rule, we get

$$\partial_t u = -\lambda g'(\xi), \quad \partial_x u = g'(\xi), \quad \partial_{xx} u = g''(\xi). \tag{23}$$

Plugging these terms in Eq. (21), we get

$$-\lambda g'(\xi) + g(\xi)g'(\xi) - \nu g''(\xi) = 0. \tag{24}$$

Taking integration with respect to ξ gives

$$-\lambda g + \frac{1}{2}g^2 - \nu g' = C, \quad C : constant. \tag{25}$$

Rewriting Eq. (25) by

$$(g - g_1)(g - g_2) = g^2 - 2\lambda g - 2C = 2\nu\, dg/d\xi, \tag{26}$$

it follows that $g_{1,2} = \lambda \pm \sqrt{\lambda^2 + 2C}$. Supposing that g_1,g_2 are real implies $g_1 > g_2$. Using separation of variable and then integrating equation (26), we get

$$g(\xi) = \frac{g_1 + g_2 e^{\left(\frac{g_1 - g_2}{2\nu}\right)\xi}}{1 + e^{\left(\frac{g_1 - g_2}{2\nu}\right)\xi}} = \frac{g_1 + g_2}{2} - \frac{g_1 - g_2}{2} \tan h\left(\frac{g_1 - g_2}{4\nu}\xi\right) \tag{27}$$

As a result the explicit form of traveling wave solution of Eq. (21) becomes

$$u(t,x) = \lambda - \frac{g_1 - g_2}{2} \tan h\left(\frac{1}{4\nu}(g_1 - g_2)(x - \lambda t)\right) \tag{28}$$

where $\lambda = \frac{g_1 + g_2}{2}$ is the wave speed. We can observe that $\lim_{\xi \to -\infty} g(\xi) = g_1$ and $\lim_{\xi \to \infty} g(\xi) = g_2$ with $g'(\xi) < 0$ for all ξ. This means the solution $g(\xi)$ decreases monotonically with ξ from the value g_1 to g_2. At $\xi = 0$, $u = \frac{g_1 + g_2}{2} = \lambda$, that is the wave form $g(\xi)$ travels from left to right with speed λ equal to the average value of its asymptotic values. The solution resembles to a shock form as it connects the asymptotic states g_1 and g_2. Without the viscosity term, the solutions to Burgers equation allow shock forms which finally break. The diffusion term prevents incrementally deformation of the wave and its breaking by withstanding the nonlinearity. As a conclusion, there exists a balance between nonlinear advection term and the linear diffusion term. The wave form is notably affected by the diffusion coefficient ν. If ν is smaller, then the transition layer between two asymptotic values of solution is sharper. In the limit $\nu \to 0$, the solutions converge to the step shock wave solutions to the inviscid Burgers' equation.

Remark. If the initial data is smooth and very small, then the u_{xx} term is negligible compared to other terms before the beginning of wave breaking. As the wave breaking starts, the u_{xx} term raises faster than u_x term. After a while, the term u_{xx} becomes comparable to the other terms so that it keeps the solution smooth, giving rise to avoid breakdown solutions.

1.6 Inviscid Burgers' equation

Whenever $\nu = 0$, Eq. (21) is called the inviscid Burgers' equation. This equation can be obtained by substituting $f(u) = u^2/2$ in the scalar conservation law (1), that is

$$\partial_t u + \partial_x (u^2/2) = \partial_t u + u\partial_x u = 0. \tag{29}$$

Observe that $f(u)$ is a nonlinear function of u; thus, the inviscid Burgers' equation is a nonlinear equation. Equation (29) is now equivalent to Eq. (17) with $\lambda = u$. We know the solution of Eq. (17); so, plugging $\lambda = u$ into the relation (20) implies that the solution of Eq. (29) is

$$u(t,x) = f(x - ut) = u_0(x - ut). \tag{30}$$

Recall that the characteristic speed λ is constant for linear advection equation; that is, the characteristic curves become parallel for Eq. (17). In contrast, for the inviscid Burgers' equation (29), the characteristic speed $\lambda = u$ depends on u. As a result the characteristic lines are not parallel. If we apply the implicit function theorem to Eq. (29), the solution can be written as a function of t and x as u_0 is differentiable. More particularly, differentiating Eq. (30) with respect to t, we get

$$\partial_t u = -u_0'(u_t t + u) \quad \Rightarrow \partial_t u = -\frac{u_0' u}{1 + u_0' t}; \tag{31}$$

and differentiating equation (30) with respect to x, we get

$$\partial_x u = u_0'(1 - u_x t) \quad \Rightarrow \partial_x u = \frac{u_0'}{1 + u_0' t}. \tag{32}$$

Thus, substituting Eqs. (31) and (32) in (29), we can recover the inviscid Burgers' equation. Consequently, the relations (31) and (32) imply that the solutions of Eq. (1) and particularly of Eq. (29) depend on the initial value u_0. It can be observed that whenever $u_0'(x) > 0$, then by Eq. (32), $\partial_x u$ decreases in time because $1 + u_{0'} t > 0$ for $t > 0$. In other words, the profile of the wave flattens as time increases. On the other hand, whenever $u_0'(x) < 0$, then $\partial_x u$ increases in time as $1 + u_0 t < 0$. Hence u_x in Eq. (32) tends to ∞ as $1 + u_0' t$ approaches to zero. As a result, wave profile become sharp after some time. For further details on the Burgers' equations, we refer the reader to [12, 13, 22] and the references therein.

1.7 Shock waves

Let the constants u_L and u_R are given with a linear function, $\varphi(t) = \lambda t$. Then

$$u(t,x) = \begin{cases} u_R & \text{if} \quad x > \lambda t, \\ u_L & \text{if} \quad x < \lambda t, \end{cases} \tag{33}$$

is a simple example of discontinuous solution of the conservation law (11). If $u_L \neq u_R$, the relation (33) is called a shock wave connecting u_L to u_R with shock speed λ. As an example, if we take into account the characteristics of the inviscid Burgers' equations which are of the form $\frac{dx}{dt} = u(t,x)$, it follows that

$$x(t) = u_0(x_0)t + x_0 \tag{34}$$

where $u_0(x) = u(0,x)$ and $x_0 = x(0)$; thus, the characteristics are straight lines. Depending on the behavior of these characteristics, we have two cases. If $u_L > u_R$, characteristics intersect, the solution will have an infinite slope, and the wave will break; as a result a shock is obtained. This is illustrated in **Figure 1**. On the other hand, if $u_R > u_L$, the characteristics do not intersect, and hence a region without characteristic will appear which is physically unacceptable. This is shown in **Figure 2**. We get rid of this by introducing the rarefaction waves.

1.8 Rarefaction waves

A rarefaction wave is a strong solution which is a union of characteristic lines. A rarefaction fan is a collection of rarefaction waves. These waves are constant on the characteristic line $x - x_0 = \alpha t$. Here $\alpha \in \left(f'(u_L), f'(u_R)\right)$ where u_L and u_R are the values of u at the edge of the rarefaction wave fan. If moreover f' is invertible, then the solution $u = u(t,x)$ satisfies

$$u(x,t) = \left(f'\right)^{-1}\left(\frac{x - x_0}{t}\right). \tag{35}$$

If, for instance, f is convex, then the rarefaction waves are increasing. If we consider again the inviscid Burgers' equation with the initial values, then the region without characteristics in **Figure 2** will be covered by rarefaction solution which is described by

$$u(t,x) = \begin{cases} u_L & \text{if} \quad x/t \leq f'(u_L), \\ \left(f'\right)^{-1}(x/t) & \text{if} \quad f'(u_L) \leq x/t \leq f'(u_R), \\ u_R & \text{if} \quad f'(u_R) \leq x/t. \end{cases} \tag{36}$$

Figure 1.
For the initial value $u_L > u_R$, characteristics, and shock wave.

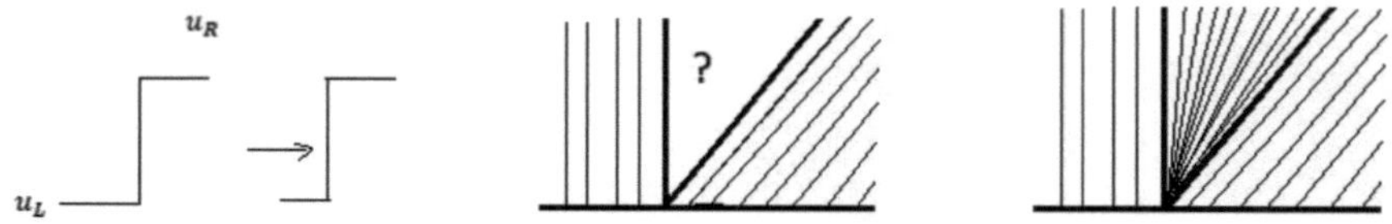

Figure 2.
For the initial value $u_R > u_L$, characteristics and rarefaction waves.

An illustration of rarefaction waves and rarefaction fan in Eq. (36) is given in **Figure 3**.

Remark. Whenever characteristics intersect, we may have multiple valued solution or no solution; but we have no more classical (strong) solution. To get rid of this situation, we introduce a more wide-ranging notion of solution, the weak solution, in the next part. By this arrangement, we may have non-differentiable and even discontinuous solutions.

1.9 Weak solution

Weak solutions occur whenever there is no smooth (classical) solution. These solutions may not be differentiable or even not continuous. Considering $\phi : \mathbb{R} \times \mathbb{R}^+ \to \mathbb{R}$ as a smooth test function with a compact support and multiplying the scalar conservation law (1) by this test function ϕ, it follows after integration by parts that

$$\begin{aligned}
&\int_0^\infty \int_{-\infty}^\infty \phi \partial_t u + \phi \partial_x f(u)\, dx\, dt \\
&= \int_{-\infty}^\infty \phi u \Big|_0^\infty dx - \int_0^\infty \int_{-\infty}^\infty u \partial_t \phi\, dx\, dt + \int_0^\infty \phi f(u)\Big|_{-\infty}^\infty dt - \int_0^\infty \int_{-\infty}^\infty f(u)\, \partial_x \phi\, dx\, dt \\
&= -\int_0^\infty \int_{-\infty}^\infty u \partial_t \phi\, dx\, dt - \int_0^\infty \int_{-\infty}^\infty f(u)\, \partial_x \phi\, dx\, dt - \int_{-\infty}^\infty u\phi \Big|_{t=0} dx.
\end{aligned} \tag{37}$$

Putting the initial condition $u_0(x) = u(0, x)$ to the above relation, it follows that

$$\int_0^\infty \int_{-\infty}^\infty u\phi_t + f(u)\phi_x dx\, dt + \int_{-\infty}^\infty u(0, x)\phi(x) dx = 0. \tag{38}$$

Observe that there are no more derivatives of u and f which may lead less smoothness. In other words, the smoothness requirement is reduced for finding a solution. Thus, the function $u(t, x)$ is said to be the weak solution of the initial value problem (11) if the relation (38) satisfied for all test function ϕ. Here it is significant to note that u needs not be smooth or continuous to satisfy Eq. (38). Consequently, by weak solutions, we extend the solutions so that discontinuous solutions may also be covered. However, in general weak solutions are not unique. We can also notice

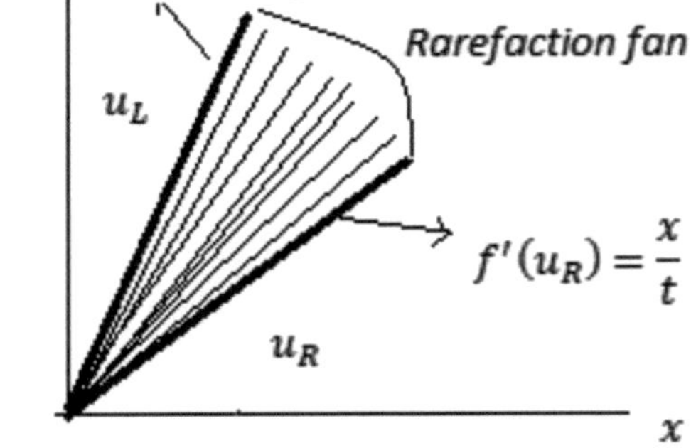

Figure 3.
Rarefaction fan.

that strong solutions are also weak solutions and a weak solution which is continuous and piecewise differentiable is also strong solution.

1.10 Riemann problem

The Riemann problem is a Cauchy problem with a particular initial value which consists a conservation law together with piecewise constant data having a single discontinuity. We consider the Riemann problem for a convex flux described by

$$\partial_t u + \partial_x(f(u)) = 0, \quad x \in \mathbb{R}, \quad t \in \mathbb{R}_+,$$
$$u(0,x) = \begin{cases} u_L & \text{if} \quad x < 0, \\ u_R & \text{if} \quad x > 0. \end{cases} \tag{39}$$

The solution is a set of shock and rarefaction waves depending on the relation between u_L and u_R. There are two cases to investigate:

Case 1: ($u_L > u_R$) A shock is obtained because the left-hand side wave moves faster than the right-hand side one. Thus the solution

$$u(t,x) = \begin{cases} u_L & \text{if} \quad x/t < \lambda, \\ u_R & \text{if} \quad x/t > \lambda, \end{cases} \tag{40}$$

is a shock wave satisfying the shock speed $\lambda = \frac{f(u_R) - f(u_L)}{u_R - u_L}$.

Case 2: ($u_L < u_R$) The solution given in Case 1 is also a solution for this case. In addition, we have rarefaction solutions of the form (36) illustrated by **Figure 3**.

1.11 Rankine-Hugoniot jump condition

A jump discontinuity along the characteristic line is controlled by the Rankine-Hugoniot jump condition. Integrating the scalar conservation law (1) in $[x_1, x_2]$, it follows that

$$\frac{d}{dt}\int_{x_1}^{x_2} u(t,x)dx + f(u)\Big|_{x_1}^{x_2} = 0. \tag{41}$$

Suppose that there is a discontinuity at the point $x = \xi(t) \in (x_1, x_2)$ where u and u' are continuous on the $[x_1, \xi(t))$ and $(\xi(t), x_2]$, respectively. Suppose also that whenever $x_1 \to \xi(t)^-$ and $x_2 \to \xi(t)^+$, their limits exist. Next, Eq. (41) can be rewritten as

$$\frac{d}{dt}\int_{x_1}^{\xi(t)} u(t,x)dx + \frac{d}{dt}\int_{\xi(t)}^{x_2} u(t,x)dx = -(f(t,x_2) - f(t,x_1)). \tag{42}$$

By the fundamental theorem of calculus, the relations (41) and (42) yield

$$u(\xi^-,x)\xi'(t) - u(\xi^+,x)\xi'(t) + \frac{d}{dt}\int_{x_1}^{\xi(t)} u_t(t,x)dx + \frac{d}{dt}\int_{\xi(t)}^{x_2} u_t(t,x)dx. \tag{43}$$

Taking the limit whenever $x_1 \to \xi(t)^-$ and $x_2 \to \xi(t)^+$, it follows that

$$\xi'(t)(x_2 - x_1) = f(x_2) - f(x_1) \Rightarrow \lambda = \xi'(t) = \frac{f(x_2) - f(x_1)}{x_2 - x_1}. \tag{44}$$

The relation (44) is said to be the Rankine-Hugoniot jump condition. Geometrical meaning of the Rankine-Hugoniot jump condition is that the shock speed is the slope of the secant line through the points $(u_L, f(u_L))$ and $(u_R, f(u_R))$ on the graph of f.

1.12 Entropy functions

Entropy and entropy flux are defined for attaining physically meaningful solutions. If u is the smooth solution of the conservation law (1), then the relation

$$\partial_t G(u) + \partial_x F(u) = 0 \tag{45}$$

is satisfied for continuously differentiable functions G and F where the pair (G, F) is called as entropy pair so that G is entropy and F is entropy flux. If in addition u is smooth, then Eq. (45) becomes

$$G'(u)\partial_t u + F'(u)\partial_x u = 0 \tag{46}$$

which looks like to the scalar conservation law (1). Indeed, if we multiply Eq. (1) by $G'(u)$, it follows that

$$G'(u)\partial_t u + G'(u)f'(u)\partial_x u = 0. \tag{47}$$

It follows that Eqs. (46) and (47) are equivalent with $F'(u) = G'(u)f'(u)$. Here the function $u(t, x)$ is said to be the entropy solution of Eq. (1) if

$$\partial_t G(u) + \partial_x F(u) \leq 0$$

holds for all convex entropy pairs $(G(u), F(u))$.

1.13 Entropy condition

Weak solutions to conservation laws may contain discontinuities as a result of a discontinuity in the initial data or of characteristics that cross each other or because of the jump conditions which are satisfied across the discontinuities. Although the Rankine-Hugoniot jump condition is satisfied, the uniqueness of the solution may always not be guaranteed. In order to eliminate the nonphysical solutions among the weak solutions, we need an additional condition, so-called entropy condition. It is described by the following: A discontinuity propagating with the characteristic speed λ given by the Rankine-Hugoniot jump condition satisfies the entropy condition if holds.

$$f'(u_L) > \lambda > f'(u_R) \tag{48}$$

Example 1.1. The weak solutions to conservation laws need not be unique. If we write the inviscid Burgers' equation in quasilinear form and multiply by $2u$, we obtain $2u\partial_t u + 2u^2\partial_x u = 0$. In conservative form it becomes

$$\partial_t\left(u^2\right) + \partial_x\left(\frac{2}{3}u^3\right) = 0, \quad \text{with } f\left(u^2\right) = \frac{2}{3}\left(u^2\right)^{3/2}. \tag{49}$$

The inviscid Burgers' equation and Eq. (49) have exactly the same smooth solutions. But their weak solutions are different. A shock traveling speed for the inviscid Burgers' equation is $\lambda_1 = (u_L + u_R)/2$; however for Eq. (49), we have $\lambda_2 = \left(\frac{2}{3}\left(\frac{u_L^3 - u_R^3}{u_L^2 - u_R^2}\right)\right)$. That is $\lambda_1 \neq \lambda_2$ whenever $u_L \neq u_R$, and thus these two equations have different weak solutions.

Example 1.2. We first consider the initial value problem for $u_L > u_R$ given by

$$\partial_t u + \partial_x\left(u^2/2\right) = 0, \quad u_0 = \begin{cases} 1 & \text{if} \quad x \leq 0, \\ 0 & \text{if} \quad x > 0. \end{cases} \tag{50}$$

Applying the method of characteristics for $t > 0$, it follows that

$$\frac{du}{dt} = 0, \quad \frac{dx}{dt} = \begin{cases} 1 & \text{if} \quad x \leq 0, \\ 0 & \text{if} \quad x > 0. \end{cases} \tag{51}$$

Next if we integrate Eq. (51) with respect to t, we get the characteristic curves

$$x = \begin{cases} t - c & \text{if} \quad x \leq 0, \\ b & \text{if} \quad x > 0, \end{cases} \tag{52}$$

where $c > 0$ and b are constants. Due to the discontinuity at the point $x = 0$, there is no strong (classical) solution. The speed of propagation is $\lambda = \frac{u_L + u_R}{2} = 0.5$. Moreover, the weak solution for $t \leq \lambda = 0.5$ becomes

$$u(t, x) = \begin{cases} 1 & \text{if} \quad \frac{x}{t} \leq 0.5 \\ 0 & \text{if} \quad \frac{x}{t} > 0.5 \end{cases}, \tag{53}$$

which satisfies both the jump condition and the entropy condition as $u_L = 1 > u_R = 0$. The characteristic curves can be observed in **Figure 4**.

Example 1.3. We now interchange the roles of u_L and u_R of the Example 1.2 so that $u_L < u_R$ to get an initial value problem:

$$\partial_t u + \partial_x\left(u^2/2\right) = 0, \quad u_0 = \begin{cases} 0 & \text{if} \quad x \leq 0, \\ 1 & \text{if} \quad x > 0. \end{cases} \tag{54}$$

By the method of characteristics, we obtain a solution

$$u_1(t, x) = \begin{cases} 0 & \text{if} \quad \frac{x}{t} \leq 1 \\ 1 & \text{if} \quad \frac{x}{t} > 1 \end{cases} \tag{55}$$

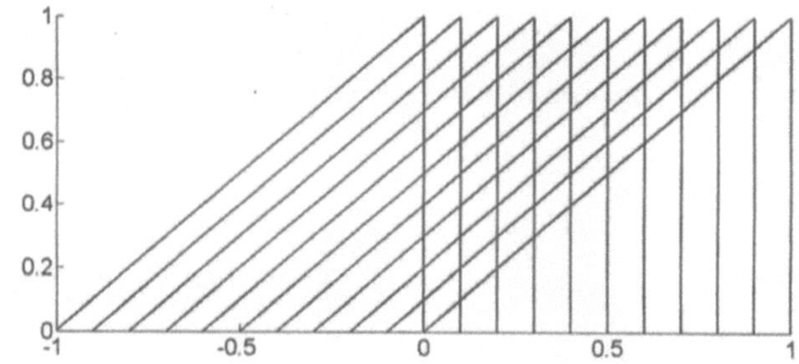

Figure 4.
For initial value $u_L > u_R$, the characteristic solutions.

Figure 5.
For initial value $u_L < u_R$, characteristic solutions $u_1(t,x)$ and $u_2(x,t)$ with rarefaction fan.

which is a classical (strong) solution on both sides of the characteristic line $\frac{x}{t} = 1$. Since it satisfies the Rankine-Hugoniot jump condition along the discontinuity curve, it is a weak solution. However, the entropy condition is not satisfied. It yields an empty region between the characteristic lines shown in **Figure 4**. In order to cover this empty state, we consider another solution described by

$$u_2(t,x) = \begin{cases} 0 & \text{if} \quad x \leq 0, \\ \dfrac{x}{t} & \text{if} \quad 0 \leq \dfrac{x}{t} \leq 1, \\ 1 & \text{if} \quad \dfrac{x}{t} \geq 1 \end{cases} \tag{56}$$

which satisfies both jump and entropy conditions. Here we can observe the rarefaction fan arising on the interval $0 \leq \frac{x}{t} \leq 1$. An illustration of this solution is supplied in **Figure 5**.

2. The gas dynamic equations in one dimension

The equation of fluid dynamics can be represented in Eulerian and Lagrangian forms. Eulerian coordinates are related to the coordinates of a fixed observer. On the other hand, Lagrangian coordinates are in usual related to the local flow velocity. That is, due to the velocity taking different values in different parts of the fluid, the change of coordinates is different from one point to another one.

2.1 Eulerian coordinates

The equations of gas dynamics in Eulerian coordinates can be written in the following conservative forms:

$$\begin{cases} \partial_t(\rho) + \partial_x(\rho u) = 0, \\ \partial_t(\rho u) + \partial_x(\rho u^2 + p) = 0, \\ \partial_t(\rho e) + \partial_x((\rho e + p)u) = 0 \end{cases} \tag{57}$$

where we ignored the heat conduction. If we denote

$$U = \begin{pmatrix} \rho \\ \rho u \\ \rho e \end{pmatrix}, \quad F(U) = \begin{pmatrix} \rho e \\ \rho u^2 + p \\ \rho e u + p u \end{pmatrix}, \tag{58}$$

then Eq. (57) can be written by

$$\partial_t U + \partial_x F(U) = 0 \tag{59}$$

where ρ is density, p is pressure, u is velocity, and e is the specific internal energy.

2.2 Hyperbolicity of the Euler system

If we do not neglect the heat conduction, then the U and F terms in Eq. (59) become

$$U = \begin{pmatrix} \rho \\ \rho u \\ E \end{pmatrix} \quad \text{and} \quad F = \begin{pmatrix} \rho u \\ \rho u^2 + p \\ (E+p)u \end{pmatrix}, \tag{60}$$

where E is total energy such that $E = \frac{1}{2}\rho u^2 + \rho e$, $e = \frac{p}{(\delta-1)\rho}$, and for perfect gases $\delta = c_p/c_v$ is the ratio of specific heats. Rewriting Eq. (59) in quasilinear form, we get

$$\partial_t U + A(U)\partial_x U = 0, \tag{61}$$

where $A(U) = \frac{\partial F}{\partial U}$ is the Jacobian matrix. The eigenvalues of $A(U)$ then are $\lambda_1 = u, \quad \lambda_2 = u - a, \quad \lambda_3 = u + a$ where a is the sound speed given by $a = \sqrt{\frac{\delta p}{\rho}}$. Moreover the corresponding eigenvectors are

$$E^{(1)} = \begin{pmatrix} 1 \\ u \\ \frac{1}{2}u^2 \end{pmatrix}, \quad E^{(2)} = \begin{pmatrix} 1 \\ u-a \\ H-ua \end{pmatrix}, \quad E^{(3)} = \begin{pmatrix} 1 \\ u+a \\ H+ua \end{pmatrix} \tag{62}$$

which are real, and the eigenvectors are linearly independent implying that the Euler equations for perfect gases are hyperbolic.

2.3 Rankine-Hugoniot conditions for the Euler system

Using the results in the previous part, the Rankine-Hugoniot jump conditions for the Euler system will be of the form

$$\begin{aligned} s(\rho_1 - \rho_2) &= m_2 - m_1, \\ s(m_2 - m_1) &= \frac{m_2^2}{\rho_2} + p_2 - \frac{m_1^2}{\rho_1} - p_1, \\ s(\rho_2 E_2 - \rho_1 E_1) &= m_2 H_2 - H_1 m_1, \end{aligned} \tag{63}$$

where the indices 1 and 2 refer to the left and right of the shock, respectively, and s denotes the wave speed.

2.4 Riemann problem for the Euler system

The Riemann problem for the one-dimensional Euler equation (57) is represented by

$$\begin{aligned} &\partial_t U + \partial_x(F(U)) = 0, \quad x \in \mathbb{R}, \quad t > 0, \\ &U(0,x) = U_0(x) = \begin{cases} U_L & \text{if} \quad x < 0, \\ U_R & \text{if} \quad x > 0. \end{cases} \end{aligned} \tag{64}$$

The reader is addressed to the references [18, 24] for further details.

2.5 Lagrangian coordinates

We aim to transform the equations of gas dynamics (57) given in the Eulerian coordinates into the Lagrangian coordinates for one-dimensional case. We start denoting by $\mathbf{u} = \mathbf{u}(t,x)$ the velocity field of the fluid flow and consider the differential system

$$\frac{dx}{dt} = \mathbf{u}(t,x). \tag{65}$$

We set the following change of coordinates from Euler coordinates to Lagrange coordinates for space and time as $(t,x) \to (t',\xi)$ where $\xi = (\xi_1,\xi_2,\xi_3) \in \mathbb{R}^3$ so that

$$t' = t, \quad \frac{\partial x(t',\xi)}{\partial t'} = \mathbf{u}(t', x(t',\xi)), \quad x_0 = x(0,\xi) = \xi. \tag{66}$$

It follows that $(t',\xi) = (t,(\xi_1,\xi_2,\xi_3))$ are the Lagrangian coordinates associated with the velocity field u. We set

$$J(t,\xi) = \det\left(\frac{\partial x_i}{\partial \xi_j}(t,\xi)\right), \tag{67}$$

which gives

$$\frac{\partial J}{\partial t}(t,\xi) = J(t,\xi)(\mathbf{div}\,\mathbf{u})(t,x(t,\xi)), \quad \text{where,} \quad \mathbf{div}\,\mathbf{u} = \sum_{j=1}^{3}\frac{\partial u_j}{\partial x_j}. \tag{68}$$

It follows by some algebraic manipulations that the gas dynamic equations become

$$\begin{cases} \partial_t(\rho J) = 0, & \text{(Conservation of mass)}, \\ \partial_t(\rho u J) + \partial_\xi(p) = 0, & \text{(Conservation of momentum)}, \\ \partial_t(\rho e J) + \partial_\xi(pu) = 0, & \text{(Conservation of energy)}. \end{cases} \tag{69}$$

In order to derive a more convenient form of the system (69), we derive firstly the equation of conservation of mass:

$$\rho J = \rho_0 = \rho(0,\xi) \tag{70}$$

where $\rho_0(\xi) = \rho(0,\xi)$. Assuming that $\rho > 0$, we introduce the specific volume $\tau = 1/\rho$, and by using Eq. (68) we get

$$J = \rho_0 \tau, \quad \text{and} \quad \partial_t J = J\partial_x u = \partial_\xi u \tag{71}$$

which yields

$$\rho_0 \partial_t \tau - \partial_\xi u = 0. \tag{72}$$

Hence the second and third equations of Eq. (69) become

$$\begin{aligned} &\rho_0 \partial_t u + \partial_\xi p = 0, && \text{(Conservation of momentum)}, \\ &\rho_0 \partial_t e + \partial_\xi (pu) = 0, && \text{(Conservation of energy)}. \end{aligned} \tag{73}$$

Moreover, we define a mass variable m by

$$m(\xi) = \int_0^{\xi} \rho(0,y)dy, \quad \text{or equivalently,} \quad dm = \rho(0,\xi)d\xi = \rho_0 d\xi. \tag{74}$$

Finally, using Eqs. (69) and (73), the Euler system (57) can be written in Lagrangian coordinates with the mass variable in the form

$$\begin{cases} \partial_t \tau - \partial_m u = 0, \\ \partial_t u + \partial_m p = 0, \\ \partial_t e + \partial_m (pu) = 0, \end{cases} \tag{75}$$

where $p = p(\tau,\xi) = p(\tau, e - u^2/2)$. If we set $V = \begin{pmatrix} \tau \\ u \\ e \end{pmatrix}$, $F(V) = \begin{pmatrix} -u \\ p \\ pu \end{pmatrix}$ with $\tau > 0, u \in \mathbb{R}, e - u^2/2 > 0,$ we obtain a scalar conservation law of the form

$$\partial_t V + \partial_m F(V) = 0 \tag{76}$$

which is strictly hyperbolic. This can be verified by checking the Jacobian matrix of the flux calculated with respect to the variables (τ, u, e)

$$\begin{pmatrix} 0 & -1 & 0 \\ p_\tau & -up_\varepsilon & p_\varepsilon \\ up_\tau & p - u^2 p_\varepsilon & up_\varepsilon \end{pmatrix} \tag{77}$$

with $e = \varepsilon + \frac{1}{2}u^2$. The eigenvalues are $\sigma_1 = -\sqrt{p_\tau - pp_\varepsilon} < \sigma_2 = 0 < \sigma_3 = \sqrt{p_\tau - pp_\varepsilon}$ so that they are all distinct, and thus the system is strictly hyperbolic.

In fact there are different versions of the gas dynamics in Lagrangian coordinates. In this part we followed the approaches stated in [9, 10, 12]. For further details we cite these works with references therein.

2.6 Rankine-Hugoniot conditions for the Lagrangian system

Similarly as in the Euler system, the Rankine-Hugoniot jump conditions for the Lagrangian system (79) are of the form

$$\begin{aligned} \sigma(\tau_1 - \tau_0) &= -(u_1 - u_0), \\ \sigma(u_1 - u_0) &= p_1 - p_0, \\ \sigma(e_1 - e_0) &= p_1 u_1 - p_0 u_0, \end{aligned} \tag{78}$$

where σ denotes the speed of propagation of the discontinuity with respect to the mass variable.

Remark. The Eulerian and Lagrangian Rankine-Hugoniot relations are equivalent. Moreover, Eulerian entropy relations are equivalent to all Lagrangian entropy relations (see [9] for further detail).

Example 2.1. For simplicity of notation, we take (t, x) as the Lagrangian coordinates. Then the system of equations

$$\begin{cases} \partial_t \tau - \partial_x u = 0, \\ \partial_t u + \partial_x p(\tau) = 0, \end{cases} \tag{79}$$

is a one-dimensional isentropic gas dynamics in Lagrangian coordinates which is also known as p-system. It is the simplest nontrivial example of a nonlinear system of conservation laws. Here τ is the specific volume, u is the velocity, and the pressure $p = p(\tau)$ is given as a function of τ by

$$p(\tau) = \kappa\tau^{-\gamma}, \quad \gamma > 0, \quad \kappa = \frac{(\gamma - 1)^2}{4\gamma}. \tag{80}$$

The system (79) is equivalent to

$$\partial_t V + \partial_x f(V) = 0, \qquad \text{with} \qquad V = \begin{pmatrix} \tau \\ u \end{pmatrix}, \quad f(V) = \begin{pmatrix} -u \\ p(\tau) \end{pmatrix}, \tag{81}$$

where $\tau > 0$ and $(\tau, u) \in \mathbb{R}^2$. If we assume that $p'(\tau) < 0$, it follows that the Jacobian matrix of f

$$J(f) = \begin{pmatrix} 0 & -1 \\ p'(\tau) & 0 \end{pmatrix} \tag{82}$$

has two real distinct eigenvalues $\sigma_1 = -\sqrt{(-p'(\tau)} < \sigma_2 = \sqrt{-p'(\tau)}$. In other words, the system (81) is strictly hyperbolic. On the other hand, for the case $p'(\tau) > 0$, it becomes elliptic. Moreover, one can verify that the solutions of the p-system (79) and the Euler system (57) are equivalent.

3. Godunov schemes

The Godunov scheme deals with solving the Riemann problem forward in time for each grid cell and then taking the mean value over these cells. The Riemann problem is solved per mesh point at each time step iteratively. If there are no strong shock discontinuities, this process may cost much and will not be effective. To get rid of such a situation, we establish approximate Riemann solvers that are easier to implement and also low cost to use. Eulerian and Lagrangian Godunov schemes are current Godunov scheme in literature. Both have advantages and disadvantages depending on the structure of the problem. A brief comparison of the method for these two approaches is presented in the last part of the chapter. In this work we will not go further in numerical examples and details of these methods; instead, we aim to present a general form of Godunov schemes for gas dynamics in Eulerian and Lagrangian coordinate. Before introducing these, we present a first-order Godunov scheme for scalar conservation laws.

3.1 First-order Godunov scheme

Consider the scalar conservation law (1). Godunov scheme is a numerical scheme which takes advantage of analytical solutions of the Riemann problem for the conservation law (1). The numerical flux functions are evaluated at the spatial steps $x_{j-1/2}$ and $x_{j+1/2}$ by handling the solutions of the Riemann problem. On each grid cell $\mathcal{I}_i = [x_{j-1/2}, x_{j+1/2}]$, we have a piecewise constant function. The Riemann problem for (1) for the left and right sides of $\mathcal{I}_i$ are described by

$$u_L(x) = \begin{cases} u_{j-1}^n & ;x < 0, \\ u_j^n & ;x > 0, \end{cases} \qquad u_R(x) = \begin{cases} u_j^n & ;x < 0, \\ u_{j+1}^n & ;x > 0, \end{cases} \tag{83}$$

respectively. These two solutions to the Riemann problem will be the numerical solution $\tilde{u}(t, x)$. Once establishing the solution over the mesh $[t^n, t^{n+1}]$, we approximate the solution at the next time step t^{n+1} by the average value

$$U_j^{n+1} = \frac{1}{\Delta x} \int_{x_{j-1/2}}^{x_{j+1/2}} \tilde{u}\left(x, t^{n+1}\right) dx. \tag{84}$$

Proceeding this process, we define the solution $\tilde{u}(x, t^{n+1})$ iteratively. Then U_j^{n+1} can be calculated by using the integral form of the conservation law (1) in the following way: We integrate (1) for $u(t, x)$ over each grid cell $[t_n, t_{n+1}] \times I_j$:

$$\begin{aligned} &\int_{x_{j-1/2}}^{x_{j+1/2}} \tilde{u}^n\left(x, t^{n+1}\right) dx - \int_{x_{j-1/2}}^{x_{j+1/2}} \tilde{u}^n(x, t^n) dx \\ &\qquad = \int_{t^n}^{t^{n+1}} f\left(\tilde{u}_{j-1/2}^n\right) dt - \int_{t^n}^{t^{n+1}} f\left(\tilde{u}_{j+1/2}^n\right) dt. \end{aligned} \tag{85}$$

Dividing both parts by Δx and using the fact that $\tilde{u}(x, t^n) = u_j^n$ is constant at the end points $x_{j-1/2}$ and $x_{j+1/2}$, we get

$$u_j^{n+1} = u_j^n - \frac{\Delta t}{\Delta x}\left(f\left(\tilde{u}_{j-1/2}^n\right) - f\left(\tilde{u}_{j+1/2}^n\right)\right). \tag{86}$$

Thus, Godunov method is a conservative numerical scheme. It can be restated in an alternative form. Assigning the constant value of u_j^n at the points $x_{j-1/2}$ and $x_{j+1/2}$ by $u^*\left(U_{j-1}^n, U_j^n\right)$ and $u^*\left(U_j^n, U_{j+1}^n\right)$, respectively, the numerical flux functions become

$$\begin{aligned} f\left(\tilde{u}_{j-1/2}^n\right) &= f\left(u^*\left(U_{j-1}^n, U_j^n\right)\right) = \mathcal{F}\left(U_{j-1}^n, U_j^n\right), \\ f\left(\tilde{u}_{j+1/2}^n\right) &= f\left(u^*\left(U_j^n, U_{j+1}^n\right)\right) = \mathcal{F}\left(U_j^n, U_{j+1}^n\right). \end{aligned} \tag{87}$$

Therefore, a first-order Godunov method takes the form

$$U_j^{n+1} = U_j^n - \frac{\Delta t}{\Delta x}\left(\mathcal{F}\left(U_j^n, U_{j+1}^n\right) - \mathcal{F}\left(U_{j-1}^n, U_j^n\right)\right). \tag{88}$$

Here the constant value of $\tilde{u}n$ depends on the initial data. In other words, the Godunov method considers the Riemann problem as constant in each grid interval $\mathcal{I}_i$. It follows that, at the subsequent time stage, the exact solutions of the problem are picked as the numerical fluxes at the grid boundary.

The Godunov method is *consistent* with the exact solution of the Riemann problem for the conservation law (1). If we suppose that $u_j^n = u_j^{n+1} = \overline{u}$, then $\tilde{u}_{j+1/2}^n = \overline{u}$ and $\mathcal{F}(\overline{u}, \overline{u}) = f(\overline{u})$. For the stability, CFL condition requires that

$$\sup_{x \in \mathcal{R}, t > 0} |f'(u(t, x))| \frac{\Delta t}{\Delta x} \leq 1 \tag{89}$$

for each u_j^n. Next, if assigning u^* as the intermediate value over the grid $\mathcal{I}_i$ in the Riemann solution, it implies that

$$u^*(u_L, u_R) = \begin{cases} u_L, & \lambda > 0, \\ u_R, & \lambda < 0, \end{cases} \tag{90}$$

where λ is the wave propagation speed. Hence the numerical flux for Godunov's method can be generalized by

$$f(u_L, u_R) = \begin{cases} \min\limits_{u_L \le u \le u_R} f(u), & \text{if} \quad u_L \le u_R, \\ \max\limits_{u_L \ge u \ge u_R} f(u), & \text{if} \quad u_R < u_L. \end{cases} \tag{91}$$

For numerical illustration of Godunov schemes, we cite the articles [14, 20, 27].

3.2 Godunov method in Eulerian coordinates

We consider Eq. (59) with (60). The eigenvalues of $F'(U)$ are $\sigma_1 = u - c < \sigma_2 = u < \sigma_3 = u + c$. Then the Riemann problem at the point $x_{i+1/2}$ between the states U_i and U_{i+1} which is solved by the Godunov scheme can be written by

$$\begin{cases} \rho_i^{n+1} = \rho_i^n - \dfrac{\Delta t}{\Delta x_i}\left((\rho u)_{i+1/2}^n - (\rho u)_{i-1/2}^n\right) \\ (\rho u)_i^{n+1} = (\rho u)_i^n - \dfrac{\Delta t}{\Delta x_i}\left((\rho u^2 + p)_{i+1/2}^n - (\rho u^2 + p)_{i-1/2}^n\right). \\ (\rho e)_i^{n+1} = (\rho e)_i^n - \dfrac{\Delta t}{\Delta x_i}\left(((\rho e + p)u)_{i+1/2}^n - ((\rho e + p)u)_{i-1/2}^n\right) \end{cases} \tag{92}$$

3.3 Godunov method in Lagrangian coordinates

Consider the initial condition for a quantity v given by the mean value

$$v_i^0 = \frac{1}{\Delta \xi_i} \int_{\xi_{i-1/2}}^{\xi_{i+1/2}} v(\xi, 0)\, d\xi. \tag{93}$$

The eigenvalues satisfy $\sigma_1 < \sigma_2 = 0 < \sigma_3$. Setting $u_{i+1/2}$ and $p_{i+1/2}$ as the values of u and p at the contact discontinuity between V_i^n and V_{i+1}^n, it follows that

$$F\left(\left(w_R\left(0; V_i^n, V_{i+1}^n\right)\right)\right) = \left(-u_{i+j/2}^n, p_{i+j/2}^n, (pu)_{i+j/2}^n\right)^T. \tag{94}$$

Then Godunov scheme for the Lagrangian coordinates takes the form

$$\begin{cases} \tau_i^{n+1} = \tau_i^n + \dfrac{\Delta t}{\Delta m_i}\left(u_{i+1/2}^n - u_{i-1/2}^n\right) \\ u_i^{n+1} = u_i^n - \dfrac{\Delta t}{\Delta m_i}\left(p_{i+1/2}^n - p_{i-1/2}^n\right) \\ e_i^{n+1} = e_i^n - \dfrac{\Delta t}{\Delta m_i}\left((pu)_{i+1/2}^n - (pu)_{i-1/2}^n\right) \end{cases} \tag{95}$$

where

$$\Delta m_i = \rho_i^0 \Delta \xi_i, \qquad p_i^n = p\left(\tau_i^n, \varepsilon_i^n\right), \quad \varepsilon_i^n = e_i^n - \frac{\left(u_i^n\right)^2}{2}. \tag{96}$$

If we now consider the moving coordinates, Godunov scheme can also be derived equivalently by the following. Setting $x_{i+1/2} = \xi_{i+1/2}$ with the approximation of $u = dx/dt$, it follows that the Eulerian coordinate $x_{i+1/2}$ of the interface $\xi_{i+1/2}$ at t_n is upgraded with respect to

$$x_{i+1/2}^{n+1} = x_{i+1/2}^n + \Delta t u_{i+1/2}^n. \tag{97}$$

Next we deduce

$$\rho_i^n \left(x_{i+1/2}^n - x_{i-1/2}^n\right) = \Delta m_i \tag{98}$$

by a simple induction process. Hence the Lagrangian Godunov schemes become

$$\begin{cases} \Delta m_i = \rho_i^0 \left(x_{i+1/2}^0 - x_{i-1/2}^0\right) \\ x_{i+1/2}^{n+1} = x_{i+1/2}^n + \Delta t u_{i+1/2}^n \end{cases} \tag{99}$$

with

$$\begin{cases} \rho_i^{n+1} = \left(x_{i+1/2}^{n+1} - x_{i-1/2}^{n+1}\right)^{-1} \Delta m_i \\ u_i^{n+1} = u_i^n - \dfrac{\Delta t}{\Delta m_i}\left(p_{i+1/2}^n - p_{i-1/2}^n\right). \\ e_i^{n+1} = e_i^n - \dfrac{\Delta t}{\Delta m_i}\left((pu)_{i+1/2}^n - (pu)_{i-1/2}^n\right) \end{cases} \tag{100}$$

Notice that the Lagrangian Godunov schemes can be reformulated as a finite volume method. Equation (100) can be written in conservative form:

$$\partial_t(\varphi J) + \partial_\xi f = 0. \tag{101}$$

If we integrate these equations on $\left(\xi_{i-1/2}, \xi_{i-1/2}\right)$ it follows that

$$\frac{d}{dt}\int_{x_{i-1/2}}^{x_{i+1/2}} \varphi d\xi + \left(f_{i+1/2} - f_{i-1/2}\right) = 0. \tag{102}$$

Here we omit the dependency of f, φ and x on t. Moreover, if we suppose that φ is constant in each cell $\left(\xi_{i-1/2}, \xi_{i-1/2}\right)$, it follows by an explicit one-step method that is

$$\Delta x_i^{n+1} \varphi_i^{n+1} = \Delta x_i^n \varphi_i^n - \Delta t\left(f_{i+1/2}^n - f_{i-1/2}^n\right). \tag{103}$$

Moreover, if (ρ, u, e) are constant in each cell with $v = u$, we get the Godunov scheme:

$$\begin{cases} \Delta x_i^n \rho_i^n = \Delta m_i \\ \Delta m_i^{n+1} u_i^{n+1} = \Delta m_i^n u_i^n - \Delta t \left(p_{i+1/2}^n - p_{i-1/2}^n \right) \\ \Delta m_i^{n+1} e_i^{n+1} = \Delta m_i^n e_i^n - \Delta t \left((pu)_{i+1/2}^n - (pu)_{i-1/2}^n \right) \end{cases} \tag{104}$$

provided $\left(u_{i+1/2}^n, p_{i+1/2}^n\right)$ are determined by the solution of the Riemann problem, which is the desired result.

3.4 Comparison of Eulerian and Lagrangian schemes

In the literature there are two types of Godunov schemes: the Eulerian and Lagrangian. To compare one with the other, both have advantages and disadvantages. These are briefly listed in the following:

3.4.1 Eulerian approach

It is more nature; that is the properties of a flow field are described as functions of the coordinates which are in the natural physical space and time. The flow is determined by examining the behavior of the functions. Eulerian coordinates correspond to the coordinates of a fixed observer. This approach is ease of implementation and computation. The computational grids derived from the geometry constraints are generated in advance. The computational cells are fixed in space, and the fluid particles move across the cell interfaces. Since the Eulerian schemes consider the implementation at the nodes of a fixed grid, this may lead to spurious oscillations for the problems like diffusion-dominated transport equations. By adding artificial diffusion, one can get rid of these oscillations; however the nature of the problem may differ from the original one. Besides, refining the grids may also lead to remove numerical oscillations, but this process may augment the computation cost. Besides, while refining the grids, it may cause restriction of the size of time step which is limited by CFL condition. This restriction does not occur in Lagrangian case.

3.4.2 Lagrangian approach

It is based on the notion of mass coordinate denoted by $m(\xi)$. An important feature of the mass coordinate is that two segments have the same length if the mass contained in these segments is the same. This leads to face with a disadvantage; that is, at each iteration time step, the problem has to be converted from the natural coordinate system to the mass coordinate system. Once the solution at the next step is known, it has to be remapped into the natural coordinate system. As a result, this process raises the cost of the computation. Lagrangian coordinates are associated to the local flow velocity. In other words, as the velocity has different values in different parts of the fluid, then the change of coordinates is different from one point to another one in Lagrangian coordinates. Thus Lagrangian coordinates are equivalent to the Eulerian coordinates at another time. Lagrangian description states the motions and properties of the given fluid particles as they travel to different locations. Hence the computational grid points are precisely fluid particles. Since the particle paths in steady flow coincide with the streamlines, no fluid particles will cross the streamlines. Hence, there is no convective flux across cell boundaries, and the numerical diffusion is minimized. As a result, Godunov method in a Lagrangian grid is easier to handle. Moreover, in the case of higher schemes, the subsonic character of the flow makes the transformation much easier than in Eulerian schemes. Lagrangian schemes consider the implementation in a grid that moves with the flow which is an advantage for the problems like the transport equations since the advective and diffusion terms can separately be examined.

Apart from the two main approaches, there is another method which is a combination of both, so-called Eulerian-Lagrangian methods. It combines the advantages and eliminates disadvantages of both approaches to get a more efficient method. For further details we address the reader to the reference in the next part.

Notes

We have tried to present only the theoretical aspects of scalar conservation laws with some basic models and provide some examples of computational methods for the scalar models. There are plenty of contributors to the subject; however, we just cite some important of these and the references therein. Scalar conservation laws are thoroughly studied in particular in [12]; for a more general introduction including systems, see [13, 15, 18, 19, 22] and the references therein. There are some important works related to the concept of entropy provided by [7, 15, 16]. A more precise study of the shock and rarefaction waves can be found in [23]. A simple analysis for inviscid Burgers' equation is done by [21]. The readers who are deeply interested in systems of conservation laws and the Riemann problem should see [8, 13, 15, 22, 24]. A well-ordered work of the propagation and the interaction of nonlinear waves are provided by [26]. We refer the reader to the papers [1, 17] for the theory of hyperbolic conservation laws on spacetime geometries and finite volume analysis with different aspects. A widely introductory material for finite difference and finite volume schemes to scalar conservation laws can be found in [18]. In this chapter we have studied the one-dimensional gas dynamics on the Eulerian and Lagrangian coordinates. For the detail on the Lagrangian conservation laws, we refer [10]; moreover for both Eulerian and Lagrangian conservation laws, we cite [11]. The proof of the equivalency of the Euler and Lagrangian equations for weak solutions is given in [25]. There are several numerical works for Lagrangian approach; some of the basic works on Lagrangian schemes are given in [2–6]. We refer the reader to the book [7] for a detailed analysis of the mathematical standpoint of compressible flows. Moreover Godunov-type schemes are precisely analyzed in [14, 27]; whereas, Lagrangian Godunov schemes can be found in [2, 12, 20]. As a last word, we must cite [9] as a recent and more general book consisting of scalar and system approaches of both Eulerian and Lagrangian conservation laws with theoretical and numerical parts which can be a basic source for the curious readers.

Author details

Baver Okutmuştur
Middle East Technical University, Ankara, Turkey

*Address all correspondence to: baver@metu.edu.tr

References

[1] Amorim P, LeFloch PG, Okutmustur B. Finite volume schemes on Lorentzian manifolds. Communications in Mathematical Sciences. 2008;**6**(4): 1059-1086

[2] Barlow AJ, Roe PL. A cell centred Lagrangian Godunov scheme for shock hydrodynamics. Computers and Fluids. 2011;**46**:133-136

[3] Carre G, Del Pino S, Despres B, Labourasse E. A cell-centered Lagrangian hydrodynamics scheme on general unstructured meshes in arbitrary dimension. Journal of Computational Physics. 2009;**228**: 5160-5183

[4] Cheng J, Shu CW, Zeng Q. A conservative Lagrangian scheme for solving compressible fluid flows with multiple internal energy equations. Communications in Computational Physics. 2012;**12**(5):1307-1328

[5] Clair G, Despres B, Labourasse E. A new method to introduce constraints in cell-centered Lagrangian schemes. Computer Methods in Applied Mechanics and Engineering. 2013; **261-262**:56-65

[6] Claisse A, Despres B, Labourasse E, Ledoux F. A new exceptional points method with application to cell-centered Lagrangian schemes and curved meshes. Journal of Computational Physics. 2012;**231**(11): 4324-4354

[7] Courant R, Friedrich KO. Supersonic Flow and Shock Waves. New York: Springer-Verlag, Interscience Publishers; 1948

[8] Dafermos CM. Hyperbolic Conservation Laws in Continuum Physics, Grundlehren der mathematischen Wissenschaften. Vol. 325. Berlin Heidelberg: Springer-Verlag; 2000

[9] Despres B. Numerical Methods for Eulerian and Lagrangian Conservation Laws. Birkhauser; 2017

[10] Despres B. Lagrangian systems of conservation laws. Numerische Mathematik. 2001;**89**:99-134

[11] Despres B. Lois de Conservation Euleriennes et Lagrangiennes, Mathematiques et Applications. Berlin Heidelberg: Springer-Verlag; 2009

[12] Godlevski E, Raviart PA. Numerical Approximation of Hyperbolic Systems of Conservation Laws, Applied Mathematical Sciences. Vol. 118. New York: Springer-Verlag; 1996

[13] Godlevski E, Raviart PA. Hyperbolic Systems of Conservation Laws. Paris, France: Ellipse; 1991

[14] Guinot V. Godunov–Type Schemes: An Introduction for Engineers. 1st ed. Amsterdam, Netherlands: Elsevier; 2003

[15] Lax PD. Hyperbolic systems of conservation laws and the mathematical theory of shock waves. In: Conf. Board. Math. Sci. Regional Conferences series in Applied Math., vol. 11. SIAM, Philadelphia; 1972

[16] LeFloch PG. Hyperbolic Systems of Conservation Laws: The Theory of Classical and Nonclassical Shock Waves. ETH Lecture Notes Series, Birkhauser; 2002

[17] LeFloch PG, Okutmustur B. Hyperbolic conservation laws on spacetimes. A finite volume scheme based on differential forms. Far East Journal of Mathematical Sciences. 2008; **31**:49-83

[18] LeVeque RJ. Finite Volume Methods for Hyperbolic Problems. 1st ed. Cambridge, England: Cambridge University Press; 2002

[19] Liu TP. Nonlinear stability of shock waves for viscous conservation laws. AMS Memoirs, 328, Providence; 1985

[20] Munz D. On Godunov-type schemes for Lagrangian gas dynamics. SIAM Journal on Numerical Analysis. 1994;**31**(1):17-42

[21] Oyar N. Inviscid Burgers equation and its numerical solutions [Master thesis]. Ankara, Turkey: METU; 2017

[22] Serre D. Systems of Conservation Laws 1–2. Cambridge, England: Cambridge University Press; 1999

[23] Smoller J. Shock Waves and Reaction–Diffusion Equations. New York: Springer-Verlag; 1967

[24] Toro EF. Riemann Solvers and Numerical Methods in Fluid Dynamics. A Practical Introduction. 3rd ed. Berlin Heidelberg: Springer-Verlag; 2009

[25] Wagner DH. Equivalence of the Euler and Lagrangian equations of gas dynamics for weak solutions. The Journal of Differential Equations. 1987; **68**:118-136

[26] Whitham G. Linear and Nonlinear Waves. New York: Wiley-Interscience; 1974

[27] Van Leer B. On the relation between the upwind-differencing schemes of Godunov, Engquist-Osher and Roe. SIAM Journal on Scientific and Statistical Computing. 2012;**5**:1-20

Chapter 5

Unsteady CFD with Heat and Mass Transfer Simulation of Solar Adsorption Cooling System for Optimal Design and Performance

Wahiba Yaïci and Evgueniy Entchev

Abstract

The purpose of the work described here was to investigate the effects of design and operating parameters on the performance of an adsorption cooling system. An unsteady Computational Fluid Dynamics (CFD) coupled with heat a mass transfer model was created for predicting the flow behaviour, pressure, temperature, and water adsorption distributions. Silica gel and zeolite 13X were both considered as possible adsorbents, though the study included silica gel given the lower working temperature range required for operation, which makes it more appropriate for residential cooling applications powered by solar heat. Validation of the unsteady computation results with experimental data found in the literature has shown a good agreement. Different computation cases during the desorption process were simulated in a parametric study that considered adsorbent bed thickness (l_{bed}), heat exchanger tube thickness (b), heat transfer fluid (HTF) velocity (v), and adsorbent particle diameter (d_p), to systematically analyse the effects of key geometrical and operating parameters on the system performance. The CFD results revealed the importance of v, l_{bed} and d_p while b had relatively insignificant changes in the system performance. Moreover, the coupled CFD with heat and mass transfer model is suitable as a valuable tool for simulating and optimising adsorption cooling systems and for predicting their performance.

Keywords: CFD analysis, adsorption cooling, solar thermally activated chiller, silica gel, zeolite, fluid flow, heat transfer, mass transfer, design, performance

1. Introduction

Conventional vapour compression cooling systems are major consumers of electricity. In addition, these systems use non-natural refrigerants, which have high global warming as well as ozone layer depletion potentials and are responsible for the emission of CO_2 and other greenhouse gases such as chlorofluorocarbons (CFCs) and hydrochlorofluorocarbons (HCFCs). From this perspective, interest in adsorption systems powered by solar energy or waste heat has been increased as they do not use ozone-depleting substances as the working fluid nor do they need electricity or fossil fuels as driving sources. Furthermore, adsorption cooling systems have many other advantages, such as simple construction, no solution pumps, powered directly by solar energy or waste heat and no need for electricity [1–3].

Adsorption is the adhesion of atoms, ions, molecules of gas, liquid, or dissolved solids to a surface. This process creates a film of the adsorbate on the surface of the adsorbent. The desorption is the reverse of adsorption. It is a surface phenomenon. The adsorption process is usually considered as physisorption, specific of weak van der Waals forces or chemisorption, specific of covalent bonding. Adsorption is generally stated by way of isotherms, which represent the quantity of adsorbate (vapour or liquid phase) on the adsorbent (solid phase) as a function of its concentration (if liquid) or pressure (if gas) at constant temperature. For convenience, the adsorbed phase is normalised by the amount of the adsorbent in order to facilitate comparison of various adsorbent-adsorbate pairs [4–6].

An adsorption cooling system consists of adsorbing material (adsorbent) packed in a vessel (adsorber) and an evacuated vessel (the evaporator). The working fluid, generally water, is the adsorbate. The working principle of the system consists in adsorbing the vapour produced in the evaporator, generating a cooling effect. Water constantly evaporates at low pressure, cooling the process air while the heat produced simultaneously in the adsorption process is removed by the cooling water, from the adsorber. At the end of the adsorption process, the desorption stage begins by heating the adsorber, using hot water, synthetic oil or any appropriate means. The extracted vapour is directed to the condenser and eventually returns as liquid to the evaporator. The thermodynamic cycle of the complete process is therefore ended. As an energy cost-effective solution, hot water/oil can be heated by free solar energy or waste heat. The two adsorption/desorption chambers of the adsorption cooling systems operate alternatively so that to generate continuous cooling power [7].

Numerous heat-pumping, refrigeration and desalination applications have been studied using various adsorbent and adsorbate pairs. Most of the cycles need medium and/or high temperature heat sources to work as the powering sources. But adsorption cycles using the silica gel/water and zeolite/water (adsorbent-adsorbate) pairs, exhibit a distinctive benefit above other systems in their capability to be driven by heat of quite low, near-ambient temperatures, so that heat from solar panels or waste heat below 100°C can be recovered, which is highly desirable, especially if flat plat collectors are used [8–11].

Nevertheless, conventional vapour compression systems still dominate in practically all applications, because adsorption cooling has some drawbacks, which require to be improved. The recognised limitation of adsorption cycles is that the heat and mass transfer coefficients of the bed are relatively small due to low conductivity of adsorbent pellets/particles and high contact resistance between particles and metal tubes/fins. The performance of an adsorbent bed is affected adversely by the heat and mass transfer limitations inside the bed, such as reduced thermal conductivity of the solid adsorbent, and internal (intra-particle) and external (inter-particle) mass transfer resistances. The other adverse consequences on the performance are: (a) lengthy adsorption/desorption cycle time; (b) low coefficient of performance (COP), resulting to enlarged energy consumption and expenditure; and (c) small specific cooling power (SCP), resulting to a bulky and outsized system [11–13].

To simulate and optimise the performance of adsorptive heat pump/cooling systems, various numerical models and several approaches have been proposed and reported in the literature.

For example, Yong and Sumathy [14] reviewed various categories of mathematical models, used to predict the functioning and effectiveness of adsorption cooling cycles. The models were sorted into three main groups: lumped parameter, thermodynamics, and heat and mass transfer models. Among the various models existing in the literature, Computational Fluid Dynamics (CFD) models based on heat and mass transfer are especially important, as they provide understanding into the operation dynamics of the adsorber in the related cooling system. A heat and mass transfer model solving the problem in the form of partial differential equations is featured, with the temperature or the mass content of adsorbate varies with space and time. Based on the geometry of the adsorption cooling system, the models can be clustered under one-dimensional (1D), two-dimensional (2D) and three-dimensional (3D) models. In general, heat and mass transfer processes are not taken into account in thermodynamic models. While heat transfer is reflected in the lumped parameters model, the temperature variation with space is not considered in it (zero-dimensional (0D) model), which is considered in the heat and mass transfer model. The distinction between the diverse models applied to simulate adsorptive cooling systems, usually relates with the variations in the simplifying assumptions, numerical solution methods, design and utilisation of the modelled system [14, 15].

Some of the previous models, which may be categorised either as uniform-temperature models [16–18] or as uniform-pressure models [19–21], considered only heat transfer while neglecting mass transfer in the adsorbent. Hajji and Worek [22] suggested a model allowing for only convection term for heat flow to assess the effect of design and operating parameters on the system performance of a zeolite heat pump system. Alam et al. [23] investigated the design parameters of the fluid side on the system performance by developing a two-dimensional heat equation for both the fluid and adsorbent sides. They used the model to explore the effect of heat exchanger design parameters on the system performance of a two-bed silica gel/water adsorption cooling unit, as well as the effect of switching speed on the system performance.

It is only lately that a number of numerical studies with consideration of using CFD and coupled heat and mass transfer have been presented. Using Darcy's law to account for the mass transfer resistance among adsorbent pellets, Sun et al. [24] studied the momentum and heat transfer in an adsorption cooling system with two adsorbent-adsorbate pairs. Their findings suggested that for low-density adsorbates, like water or methanol, the operation dynamics of the adsorber may possibly be critically controlled by mass transfer resistance within the adsorption unit if its size is bulky. Their model is 1D. Amar et al. [25] analysed a 2D model, which similarly took into account the combined heat and mass transfers in the adsorber to investigate the impacts of different functional parameters on the performance of a thermal wave regenerative heat pump. A three-dimensional model was investigated by Zhang and Wang [11] and Zhang [12] to study the effect of coupled heat and mass transfer in adsorbent beds on the performance of a waste heat adsorption cooling unit. They also studied the effect of reactor configuration on the performance. Solmus et al. [26, 27] presented numerical studies of heat and mass transfer within the adsorbent bed of a silica gel/water adsorptive cooling system by means of the local volume averaging method. They utilised a transient 1D local thermal non-equilibrium model, which taken into account both internal and external mass transfer resistances. Their results showed the significance of spatial temperature and pressure gradients clearly indicated that external mass transfer resistance and heat transfer were important. Caglar et al. [28] developed a 2D mathematical model of the heat and mass transfer inside a cylindrical adsorption bed for a thermal wave adsorption cycle, with a heat transfer fluid flowing through an inner tube and the adsorbent in the annulus. He investigated the effect of determining factors that enhance the heat and mass transfer inside the adsorbent bed. Çağlar [29] used a 2D coupled heat and mass transfer model to analyse both finless and finned tube-type adsorbent bed for a thermal wave adsorption cooling cycle with silica gel/water as the working pair. He showed that a significant enhancement in the heat transfer is obtained using a finned tube such that the temperature of the adsorbent in the finned tube adsorbent bed.

Despite the continuous research effort about the dynamic behaviour of heat and mass transfers inside the adsorbent bed and attempts to enhance the overall system performance, there remains comprehensive research effort to be made for the accurate design and performance prediction of adsorption cooling systems. Although the limited number of studies have dealt recently with multi-dimensional effects, most of the modelling efforts have focused on a one-dimensional description of the adsorption process for its simplicity, either on adsorption or desorption processes. However, 0D or 1D model cannot describe the flow structure, the dynamic behaviour and interactions of heat and mass transfers inside the adsorbent bed and the heat exchanger. In addition, in the analyses, by assuming an equilibrium adsorption state, the internal mass transfer resistance is not taken into account, which can be very limiting for performance enhancement when the cycle time is small.

In the present work, the configuration and operating conditions influence on the performance of a solar heat driven adsorption cooling system operating in desorption mode is simulated. A 3D/2D unsteady CFD coupled with heat and mass transfer model using silica gel/water or zeolite/water pairs is created, and validated from literature data. Effects of the adsorption bed with a finned tube heat exchanger geometry as well as the operating conditions on the system performance are then fully investigated in detail. Distinct characteristics of significance to optimum design and operation that have effects on the adsorption cooling system are demonstrated and analysed. The CFD model developed in this study may be useful to design and optimise a new and more efficient adsorption cooling bed. It also provides a tool for optimisation of adsorption cooling systems driven by solar heat or low-grade/waste heat.

For this purpose, the rest of the paper is organised as follows. In Section 2, in depth steps taken for developing the CFD coupled with heat and mass transfer model of a solar adsorption cooling system, including the physical model, the governing equations, boundary conditions, and the numerical procedure are thoroughly described. The validation of the model is provided in Section 3. Section 4 discusses in detail the base case simulation and the parametric study results of the solar-driven adsorption cooling system. Finally, Section 5 summarises the main conclusions of the work.

2. CFD model details

2.1 Physical model

A heat exchanger in adsorption cooling system is a device that is in thermal contact with the adsorbent (solid phase) and helps to heat and cool the adsorbent throughout the desorption and adsorption periods, respectively. The plate-finned bed can take various configurations, such as the finned tube type, the fin plate type, and the flat-pipe type [15, 30]. A basic design of a finned tube heat exchanger is displayed in **Figure 1**.

From the possible adsorber bed geometries, the finned tube was selected to study. **Figure 2a** and **b** present the geometric model of the finned tube adsorption cooling unit. **Figure 2a** shows the full geometry consisting of a single copper tube

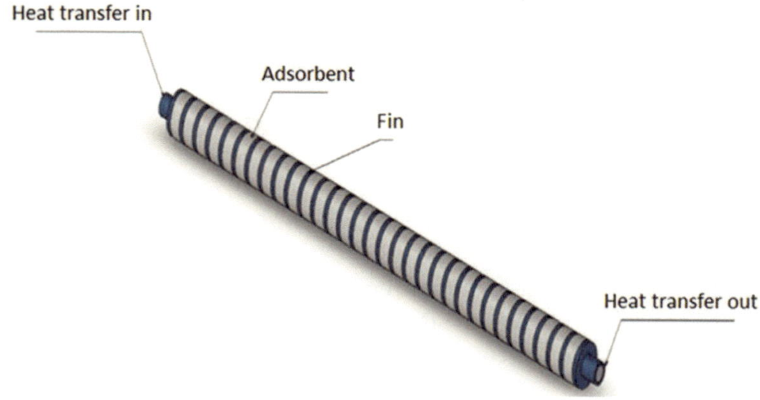

Figure 1.
A simple annular-finned tube used in adsorption cooling system [15].

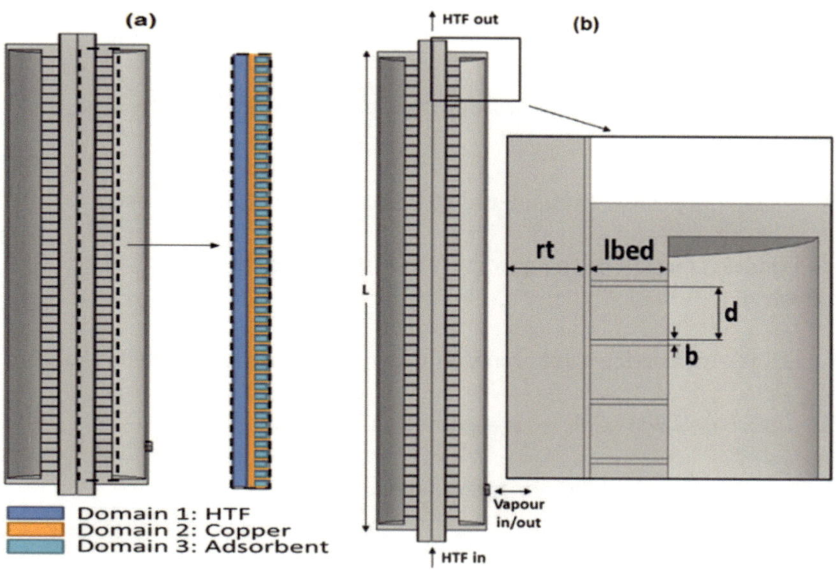

Figure 2.
Simulated finned tube adsorber: (a) geometric selection and domains and (b) schematic of the geometry and nomenclature.

with fins extending off of the tube. The HTF flows in though the bottom of the tube and exits through the top to control the temperature of the unit. Around the tube is the adsorbent material that is secured between the copper fins. There are a total of 38 fins in the full geometry. The system is symmetrical, thus the computational domain is modelled with symmetric boundary conditions; the geometry was reduced a 2D axisymmetric face of the 3D geometry, as visualised in **Figure 2a**. **Figure 2b** depicts the schematic of the model with fully labelled geometric lengths, which will be investigated in the parametric study.

Unsteady flow models in desorption mode were built in the commercial software Multiphysics COMSOL [29]. It was used as the grid generator and as the CFD solver. In order to better predict the different field characteristics, the optimised solution-adaptive mesh refinement is used. More cells were added at locations where significant phenomena changes are expected, for example near the adsorber/heat exchanger walls and inlet/outlet ports. The resulting mesh thus enabled the features of the different fields to be better resolved. The symmetric solver selected here accounts for the three-dimensional effects. Mixed topology of unstructured grids was utilised, and the final mesh was composed of about 35,305–85,000 elements depending on the volume of the finned tube adsorption bed considered. Details on the grid system and selected mesh elements can be found in Section 2.3.

2.2 Governing equations and boundary conditions

The Navier-Stokes and the mass and energy equations in three-dimensional form were used to solve for the transient fluid dynamics with the coupled heat and mass transfer fields inside the finned tube adsorption bed system with silica gel/water or zeolite/water as the adsorbent/adsorbate working pairs. The mass balance describes the rate of adsorption or desorption within the adsorbent bed. The models were created in the COMSOL Multiphysics software package including the CFD, Heat Transfer and Chemical Reaction Engineering modules [31].

The governing equations of mass, momentum and energy conservation were solved by using the finite element method, based on the following assumptions:

- the finned tube adsorption bed operates under unsteady-state, non-equilibrium conditions;
- the surface porosity is considered to be equal to the total porosity;
- the volume fraction of the gas phase is assumed to be equal to the total porosity;
- the absorbent particles are assumed to have uniform size, shape and porosity;
- the adsorbate's water phase is assumed to be a liquid;
- the adsorbate's water vapour phase is assumed to be an ideal gas;
- the adsorbed phase is considered to be a liquid, while the gas phase is assumed to be an ideal gas;
- the working fluid is water at high velocity is assumed to be Newtonian, incompressible flow inside the tubes, resulting in a constant tube-surface temperature; the viscous dissipation and viscous work are neglected; there are no body forces;
- the work done by pressure changes, radiative heat transfer and viscous dissipation is neglected; and
- the wall thickness of the vacuum tube is assumed to be very thin and hence, its thermal resistance was neglected.

Therefore, the resulting governing equations can be stated as follows. For conciseness, the basic variables defined in the following equations can be found in [12–14, 24, 26, 28].

Conservation of mass:

$$\varepsilon\frac{\partial \rho_v}{\partial t} + (1-\varepsilon)\rho_s\frac{\partial X}{\partial t} - D_m\Delta^2\rho_v + \Delta(\boldsymbol{u}\rho_v) = 0 \tag{1}$$

Conservation of momentum:

The momentum balance expresses the motion of the HTF within the tube.

$$\rho \frac{\partial \boldsymbol{u}}{\partial t} + \rho(\boldsymbol{u} \cdot \nabla)\boldsymbol{u} = \nabla \cdot \left[-pI + \mu\left(\nabla \boldsymbol{u} + (\nabla \boldsymbol{u})^2 \right) \right] \tag{2}$$

$$\rho \nabla \cdot (\boldsymbol{u}) = 0 \tag{3}$$

Conservation of energy:

The energy transfer within the system can be described with respect to the three discrete domains involved: the HTF, the metal tube and fins, and the adsorbent. The adsorbent domain is assumed to be under local thermal equilibrium between the adsorbed and vapour phase. Therefore, the equation has three additional terms to take into account for the change of temperature due to the water vapour, water adsorbed and the heat of adsorption associated with the adsorption process.

Heat transfer in the HTF:

$$\rho_f c_f \frac{\partial T_f}{\partial t} - k_f \Delta^2 T_f + \rho_f c_f v_f \Delta T_f = 0 \tag{4}$$

Heat transfer in the metal:

$$\rho_t c_t \frac{\partial T_t}{\partial t} - k_t \Delta^2 T_t = 0 \tag{5}$$

Heat transfer in the adsorbent:

$$\begin{aligned} &(1-\varepsilon)\rho_s c_s \frac{\partial T_s}{\partial t} - \varepsilon \rho_v c_{p,v} \frac{\partial T_s}{\partial t} + (1-\varepsilon)\rho_s X c_{p,l} \frac{\partial T_s}{\partial t} \\ &-(1-\varepsilon)\rho_s |\Delta H| \frac{\partial X}{\partial t} - k_s \Delta^2 T_s + \rho_v c_{p,v} \boldsymbol{u} \Delta T_s = 0 \end{aligned} \tag{6}$$

The heat of adsorption (kJ/kg) changes linearly with the amount of water adsorbed to the adsorbent [13]:

$$\Delta H = 2950 - 1400X \tag{7}$$

The velocity of the water vapour is calculated using Darcy's equation, which defines the external mass transfer resistance in the water adsorption process. Darcy's equation is valid in this condition since the adsorbent is porous and both the velocity and the pressure are relatively low.

The 2D cylindrical and 3D Cartesian coordinate expressions are both represented:

$$u = -\frac{\kappa}{\mu}\frac{\partial P}{\partial r}, v = -\frac{\kappa}{\mu}\frac{\partial P}{\partial z} \tag{8}$$

$$u = -\frac{\kappa}{\mu}\frac{\partial P}{\partial x}, v = -\frac{\kappa}{\mu}\frac{\partial P}{\partial y}, w = -\frac{\kappa}{\mu}\frac{\partial P}{\partial z} \tag{9}$$

The internal mass transfer resistance for the water adsorption process is defined by means of a linear driving force expression as follows:

$$\frac{\partial X}{\partial t} = k_m (X_e - X) \tag{10}$$

where X represents the amount adsorbed and X_e is the equilibrium adsorption capacity of the adsorbent-adsorbate pair under study, provided in Eqs. (13)–(18).

The temperature dependent mass transfer coefficient and diffusion coefficient are calculated as follows [25]:

$$k_m = \frac{15}{r_p^2} D_m \tag{11}$$

$$D_m = D_o \exp\left(-\frac{E_a}{RT_S}\right) \tag{12}$$

where D_o is reference diffusivity and r_p is the particle radius.

2.2.1 Kinetic expression for regular density (RD) silica gel

Regular density (RD) silica gel is the most widespread type of silica gel and is obtainable from any supplier that retails silica gel. Because of its capacity for high moisture uptake in the low RH (relative humidity) range, it is a very effective desiccant.

The equilibrium adsorption capacity characterises the theoretical maximum capacity that the adsorbent bed can adsorb for a given pressure and temperature [32].

For RD silica gel, the following pressure dependent isotherm is used as follows:

$$X_e = \mathrm{a}\left(\frac{\mathrm{P}}{\mathrm{P_{sat}}}\right)^{\mathrm{b}} \tag{13}$$

The values of a and b are temperatures dependent are provided in the next section, **Table 1** and given as follows [33]:

$$a = a_0 + a_1 T_s + a_2 T_s^2 + a_3 T_s^3 \tag{14}$$

$$b = b_0 + b_1 T_s + b_2 T_s^2 + b_3 T_s^3 \tag{15}$$

The saturated water pressure (kPa) for the adsorbate is provided by Antoine's equation [34].

$$P_{sat} = 0.1333 \cdot 10^{8.07131 - \frac{1730.63}{T_s - 39.724}} \tag{16}$$

2.2.2 Kinetic expression for zeolite 13X

The equilibrium adsorption capacity for zeolite 13X is determined with the following Dubinin-Astakhov (D-A) equation as the adsorption isotherm where the values of B and n are constants [35]:

$$X_e = X_o \exp\left[-B\left(\frac{T_s}{T_{sat}} - 1\right)^n\right] \tag{17}$$

The saturated water temperature (K) for the adsorbate is given by Antoine's equation [34] as follows:

$$T_{sat} = 39.724 + \frac{1730.63}{8.07131 - \log 10\left(7.500638 \cdot 10^{-3} \cdot P\right)} \tag{18}$$

Assuming symmetry conditions, the boundary conditions in a view of the 2D axisymmetric finned adsorbent, used in this work are depicted **Figure 3**. At the upstream or inlet boundary, Dirichlet boundary conditions, uniform flow with

Category	Description	Variable	Unit	Validation zeolite 13X/water	Base case silica gel/water
Temperature pressure	Hot temperature	T_{hot}	K	473	338
	Cold temperature	T_{cold}	K	313	298
	Condenser pressure	P_c	kPa	4.247	4.246
	Evaporator pressure	P_e	kPa	n/a	1.228
	Adsorption initial concentration	X_{a0}	kg/kg	—	0.024
	Desorption initial concentration	X_{d0}	kg/kg	—	0.72
Geometry	Bed thickness	l_{bed}	mm	13	12
	Fin distance	d	mm	10	15
	Tube thickness (copper)	b	mm	1	1
	Adsorbent particle diameter	d_p	mm	1.25	3
	Absorber length	L	mm	500	600
	Tube radius	r_t	mm	13	13
	Tube thickness	b_t	mm	1	1
	Void fraction	ε	—	0.635	0.635
	Permeability	κ	m^2	3.04e-7	3.4e−9
Mass	Reference diffusivity	D_o	m^2/s	5.8e-9	2.54e−4
	Activation energy of surface diffusion	E_a	J/mole	1.0e5	3.36e4
	Fluid density	ρ_f	kg/m^3	914	f(T)
	Absorbent density	ρ_s	kg/m^3	1000	670
	Tube density	ρ_t	kg/m^3	8700	8700
	Dynamic viscosity	μ	kg/m s	1.0e-5	1.5e−5
Thermal properties	Fluid velocity	v	m/s	0.001	0.001
	Thermal conductivity of bed	k_s	W/m K	1	0.3
	Thermal conductivity of fluid	k_f	W/m K	0.155	f(T)
	Thermal conductivity of tube	k_t	W/m K	400	400
	Convective heat transfer coefficient: tube and adsorbent	h_o	W/m^2 K	100	100
	Convective heat transfer coefficient: tube and fluid	h_i	W/m^2 K	100,000	100,000
	Specific heat of fluid	c_f	J/kg K	1930	f(T)
	Specific heat of particle liquid	c_{pl}	J/kg K	4180	4180
	Specific heat of particle vapour	c_{pv}	J/kg K	1880	1880
	Specific heat of adsorbent solid	c_s	J/kg K	837	f(T)
	Specific heat of tube	c_t	J/kg K	385	f(T)
Kinetics	Water adsorbed reference	X_o	kg_w/kg_s	0.261	—
	D-A constant	B	—	5.36	—
	D-A constant	n	—	1.73	—

Category	Description	Variable	Unit	Validation zeolite 13X/water	Base case silica gel/water
	Kinetic constant	a_0	—	—	−6.5314
	Kinetic constant	a_1	—	—	0.072452
	Kinetic constant	a_2	—	—	−0.00023951
	Kinetic constant	a_3	—	—	2.5493e−7
	Kinetic constant	b_0	—	—	−15.587
	Kinetic constant	b_1	—	—	0.15915
	Kinetic constant	b_2	—	—	−0.00050612
	Kinetic constant	b_3	—	—	5.329e−7
Misc.	Universal gas constant	R	J/mole K	8.314	8.314
	Ideal gas constant for water vapour	R_v	J/kg K	461.5	461.5

Table 1.
Parameter settings for the validation and parametric studies.

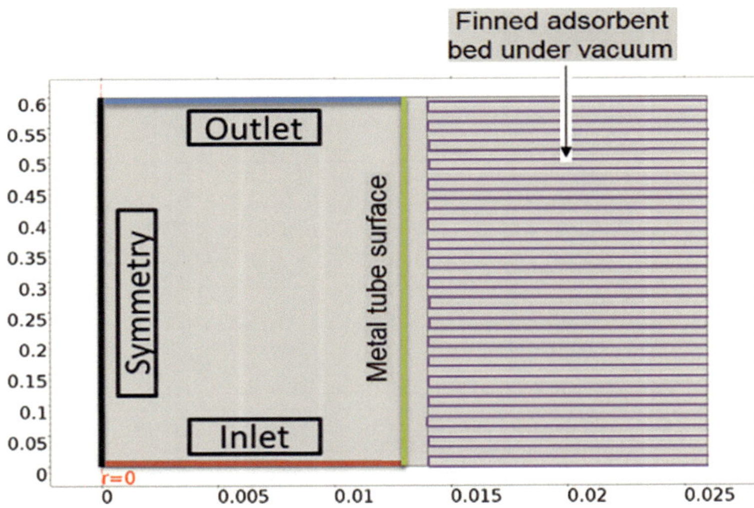

Figure 3.
View of the 2D axisymmetric finned tube adsorber: computational domain with boundary conditions.

constant velocity and constant temperature are assumed. At the downstream end of the computational domain or outlet, the Neumann boundary condition is used, i.e. stream wise gradients for all the variables are set to zero. No-slip boundary condition is used at the adsorbent fin surfaces. These surfaces are assumed to be solid walls with no slip wall boundary condition; the velocity of the fluid at the wall is zero and constant wall temperature is presumed. This sweeping statement relates to isothermal wall boundary condition. The fins and tube are presumed to be made of copper. As copper is a rather high thermal conductivity material, constant wall temperature boundary condition can be confidently supposed throughout the shells. In the right side of the figure, is shown the adsorbent bed under vacuum, where there is zero flux for mass transfer.

For energy transfer, symmetry, inlet and outlet of the metal tube, have the same definitions as in the momentum transfer physics. However, the metal tube surface with the fins shown in green and purple lines, define the heat transfer between the three domains. The convective heat transfer boundary conditions between the heat transfer fluid, metal tube and the adsorbent are defined as follows:

HTF to metal tube:

$$q_i = h_i(T_t - T_f) \tag{19}$$

where h_i is the convective heat transfer coefficient between tube and fluid.

Metal tube to adsorbent:

$$q_o = h_o(T_s - T_t) \tag{20}$$

where h_0 is the convective heat transfer coefficient between tube and adsorbent.

2.3 Numerical procedure

The governing equations of mass, momentum and energy conservation are solved by using the finite element method, based on the assumptions listed in Section 2.2. The governing equations are discretised on the computational domain, linearised in an implicit manner and solved by the finite element method using a pressure-based coupled solver (PBCS). This latter solves pressure and momentum

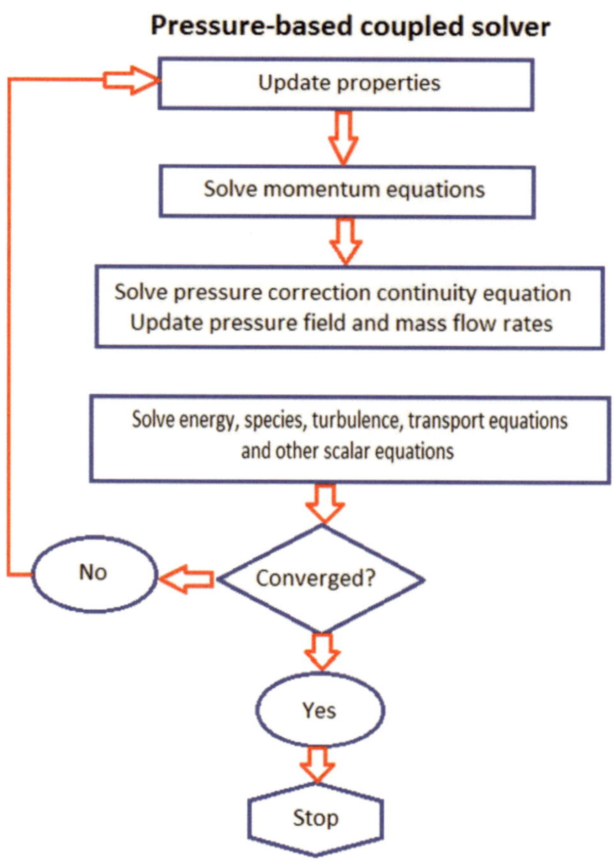

Figure 4.
Overview of the pressure-based coupled solver algorithm.

simultaneously. **Figure 4** represents an overview of the PBCS algorithm. SIMPLEC (Semi-Implicit Method for Pressure Linked Equations-Consistent) algorithm is applied for the pressure-velocity coupling; the second order upwind discretisation scheme is used for the convection terms and each governing equation is solved using QUICK (Quadratic Upwind Interpolation) scheme [36].

It is required to carry out independency verification of the grid system before CFD computation. The mesh independence study is examined by utilising three different mesh sizes (normal, fine and finest meshes) of 23,083, 35,305, and 84,992 for the finned tube adsorption cooling bed are adopted for computation for the baseline case. The relative error compared to the fine mesh, in the average bed temperature and the total water adsorbed as a function of time, are 2.85 and 1.84%, and 0.13 and 0.07%, respectively, before settling to a fine mesh for the geometry of the computational adsorption cooling bed cases. Computations were then run for a geometry comprising about 35,305 meshes, which was considered satisfactory in terms of accuracy and efficiency.

Furthermore, the solution is iterated until convergence is achieved, that is, residual for each equation achieves values less than 10^{-6}, and variations in energy, mass and temperature, respectively become negligible. A workstation with 2 (R) Xeon processors and a 2 core 2.4 GHz CPU with an installed memory of 32 GB (RAM), which took between 4 and 8 hours of CPU time depending on the computation case, was utilised to execute the necessary task.

3. Model validation

Experimental data of a finned tube adsorption bed with zeolite/water (adsorbent/adsorbate) as the working pair, zeolite as the adsorbent material by Çağlar et al. [28] have been used to validate the CFD model in the present study. The finned tube heat exchanger geometry was modified to have the identical geometric dimensions and operating conditions as found in [27]. **Table 1** shows the parameter settings associated with the experimental operation with particular settings associated with the experimental tests for t, k, v, l_{bed} and r shown in **Figure 5**. In **Figure 4**

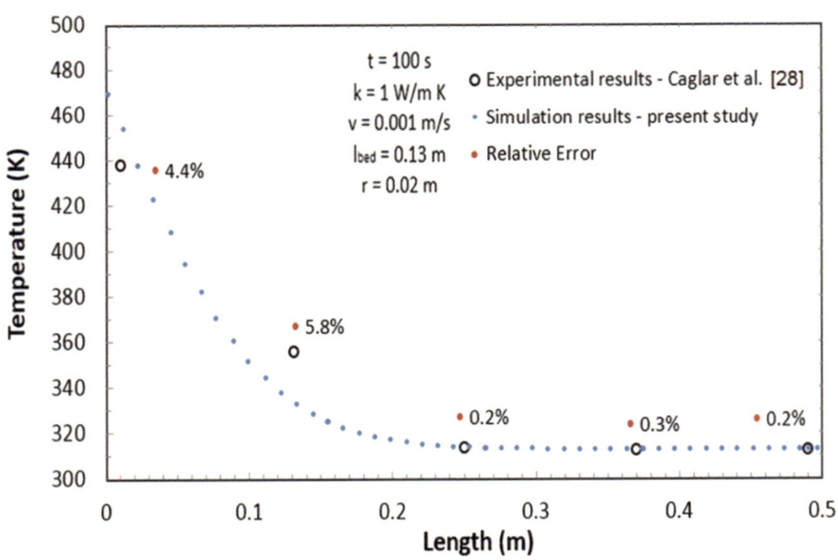

Figure 5.
Comparison of simulation results with experimental results of Çağlar et al. [28] in finned tube adsorption cooling system using zeolite/water for validation study.

shows the results of the simulated temperature profile in COMSOL Multiphysics along r = 0.02 m of the unit. It can be seen that the experimental data for the desorption mode compared to the CFD simulation results provide good agreement between the each other with deviation of no more than 5.8%. The difference between the experimental and numerical results is higher at the inlet of the unit. This may be due to the experimental uncertainties due to the temperature measurement errors because of lower heat transfer resulted from flow maldistribution or various losses. However, the observed uncertainties are well within the uncertainties of sensor measurements. This CFD model of the adsorption cooling system was therefore utilised for further transient analysis with supporting reliability of the computation.

4. Results and discussion

The selected adsorption chiller applied a finned tube geometry with a silica gel/water working pair. A base line case on desorption mode is initially simulated and studied to determine the intrinsic behaviour of the system. The parameter values for the base case and for the parametric study are presented in **Table 1**.

4.1 Base case

Figure 6 presents a 3D qualitative assessment of the pressure, temperature and water desorption distributions at operation times of 200, 600 and 1200 s. The temperature profile captures the heat transfer characteristics of the system throughout the entire system while the pressure and adsorbed water only capture phenomena in the adsorbent bed. Desorption is the mode of operation highlighted in this study. The HTF inlet temperature is 338 K while the bed temperature is 298 K. The HTF enters the tube at z = 0 and exits at z = 0.6 m. A white set of arrows

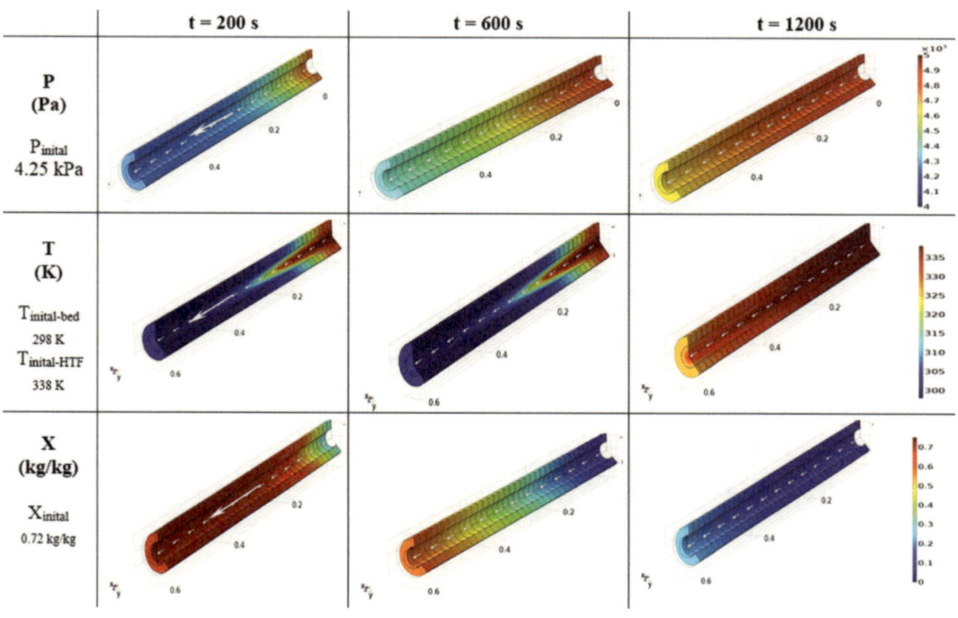

Figure 6.
Pressure, temperature and adsorbed water distributions at t = 200, 600 and 1200 s of the 3D geometry for the baseline case under desorption mode.

show the direction of the fluid flow. The kinetics of the adsorbed water is slow enough that there exists a time delay between reaching the thermodynamic temperature and establishing the equilibrium. The adsorbent is fully saturated after 2000 s, but this is not illustrated on the figure for brevity. For desorption the initial pressure of the system increases to 4.246 kPa and the initial adsorbed water begins at 0.72 kg/kg and reaches 0.08 kg/kg upon completion. The pressure of the system closely follows the behaviour of the temperature profile and affects the equilibrium water adsorption.

Figure 7 presents a 2D qualitative assessment of the pressure, temperature and desorbed water distributions. The same layout and results are chosen to be displayed as in **Figure 6** for an alternative, but more insightful, assessment of the simulation results. From this visual perspective, the profiles for all simulated measures may be better compared in the parametric study.

4.2 Parametric study

A parametric study applying the developed transient CFD coupled with heat and mass transfer model, was conducted in an effort to predict the influence of various parameters on the design and performance of an adsorption cooling system during the desorption process. The effect of these parameters on the governing independent parameters influencing the fluid flow, the heat and mass transfer on the adsorber performance, are the geometrical, particle size, physical-chemical, thermodynamic and thermal property parameters of the adsorber. The parameters investigated in this study are the adsorbent bed thickness (l_{bed}), the heat exchanger tube thickness (b), the HTF velocity (v), and the adsorbent particle diameter (d_p).

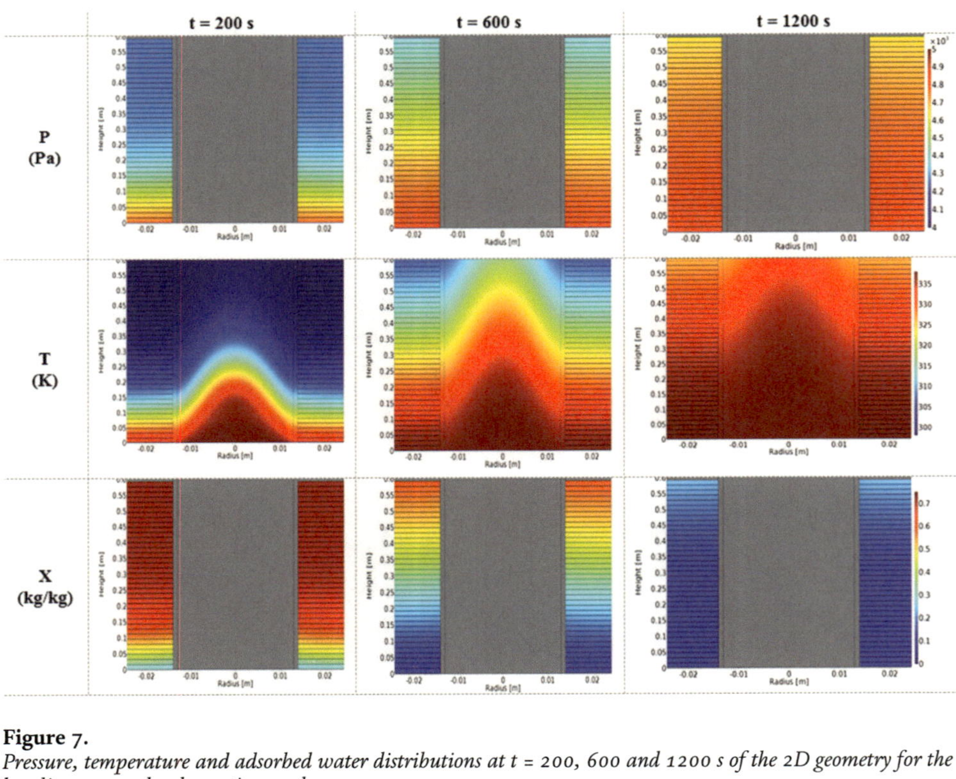

Figure 7.
Pressure, temperature and adsorbed water distributions at t = 200, 600 and 1200 s of the 2D geometry for the baseline case under desorption mode.

Case description			Geometry		Fluid flow and particle size properties	
Case	**Case no.**	**Case description**	**Bed thickness**	**Tube thickness**	**HTF velocity**	**Adsorbent particle diameter**
			l_{bed} **(mm)**	b **(mm)**	v **(m/s)**	d_p **(mm)**
1	1	Base	12	1	0.001	3
2	2 3	Bed thickness	30 45	1	0.001	3
3	4 5	Tube thickness	12	1.5 2	0.001	3
4	6 7	Fluid velocity	12	1	0.005 0.015	3
5	8 9	Adsorbent particle diameter	12	1	0.001	2 5

Table 2.
Summary of the simulation matrix with the input data of the adsorption cooling system.

The basic aim of the analysis is to better understand the optimum design of the adsorption cooling system with heat exchanger configurations using CFD approach and, thereby, learn how to successfully maximise the overall heat and mass transfer performance of the system such as temperature and desorption rates whilst minimising the pressure drop.

Table 2 depicts the four-parameter settings, which were selected in the study. Eight cases were investigated.

4.2.1 Effect of adsorbent bed thickness (l_{bed})

The effect of adsorbent bed thickness on the dependent variables pressure, temperature and amount of water adsorbed within the adsorption cooling system is investigated for adsorbent bed thicknesses of 12 (baseline case), 30 and 45 mm. **Figure 8** presents the effect of bed thickness on simulated transient average bed pressure, average bed temperature and adsorption of water profiles. **Figure 9** shows the simulated temperature and water adsorbed distributions at t = 600 s.

The results reveals that the thickness of adsorbent bed has a substantial effect on the desorption performance. As the bed thickness increases, the bed temperature takes more time to increase. Therefore, the rate of the pressure drop is lower and the water adsorption takes a greater extent of time. The pressure, temperature and the amount of water adsorbed distributions inside the bed reaches equilibrium after about 1500, 2500 and 3000 s for bed thicknesses of 12, 30 and 45 mm, respectively. The desorption bed produced decreased average pressure values of 0.8 and 1.6%, for adsorption bed thicknesses of 30 and 40 mm relative to the baseline case having a bed thickness of 12 mm, respectively. Alternatively, under the same conditions, the desorption bed produced the same amounts of decreased average temperature and decreased adsorbed water values of 15.3 and 35.0%, respectively.

As adsorbent bed thickness increases, more adsorbent is used, external mass transfer resistances increase with increasing mass of adsorbent. Thus, more heat is expected to be required to increase the temperature of the entire unit; the thermal resistances across the adsorbent bed are substantially increased. Fin temperature is reduced and cannot generate a sufficient heat transfer since the contact resistances between fins and adsorbent material increase. This causes reduced mass transfer, smaller amounts of adsorption capacities and increased cycle times. This same

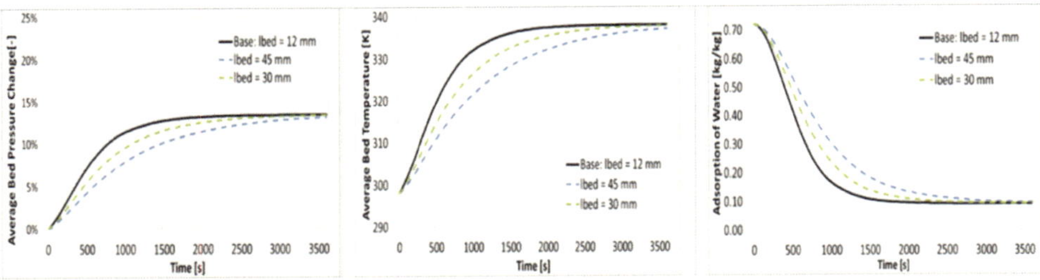

Figure 8.
Effect of bed thickness (l_{bed}) on desorption 1D profile as a function of time for average bed pressure drop, average bed temperature and adsorption of water.

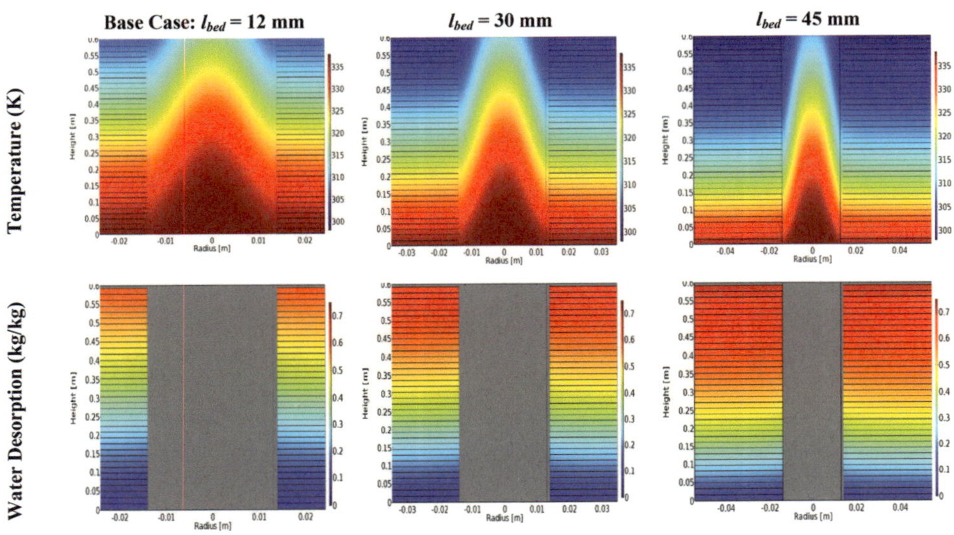

Figure 9.
Effect of bed thickness (l_{bed}) on desorption 2D temperature and adsorbed water distributions at t = 600 s.

outcome has been documented by previous researches in heat exchangers in general. Heat transfer in the radial direction is improved by reducing the adsorbent bed thickness. For design considerations, a reduced adsorbent bed thickness will produce a reduced cycle time and an enhanced specific cooling power. Increasing bed thickness has then an adverse effect on the heat transfer within the bed. There is a trade-off to consider; decreasing bed thickness, results in an improvement in the heat transfer across the adsorbent bed at the expense of a reduction in the mass of adsorbent and hence, of the adsorption capacity.

4.2.2 Effect of tube thickness (b)

The effect of tube thickness on the dependent variables pressure, temperature and amount of water adsorbed within the adsorption cooling system is examined for tube thickness of 1 (baseline case), 1.5 and 2 mm. **Figure 10** shows the effects of tube thickness on simulated transient average bed pressure, average bed temperature and adsorption of water profiles. **Figure 11** presents the simulated temperature and water adsorbed distributions at t = 600 s.

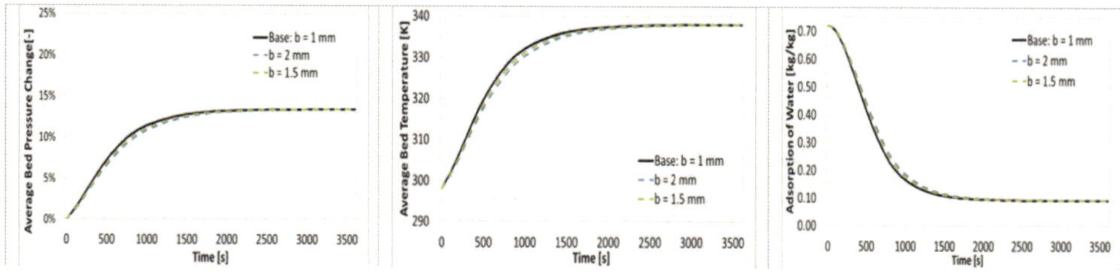

Figure 10.
Effect of tube thickness (b) on desorption 1D profiles as a function of time for average bed pressure drop, average bed temperature and adsorption of water.

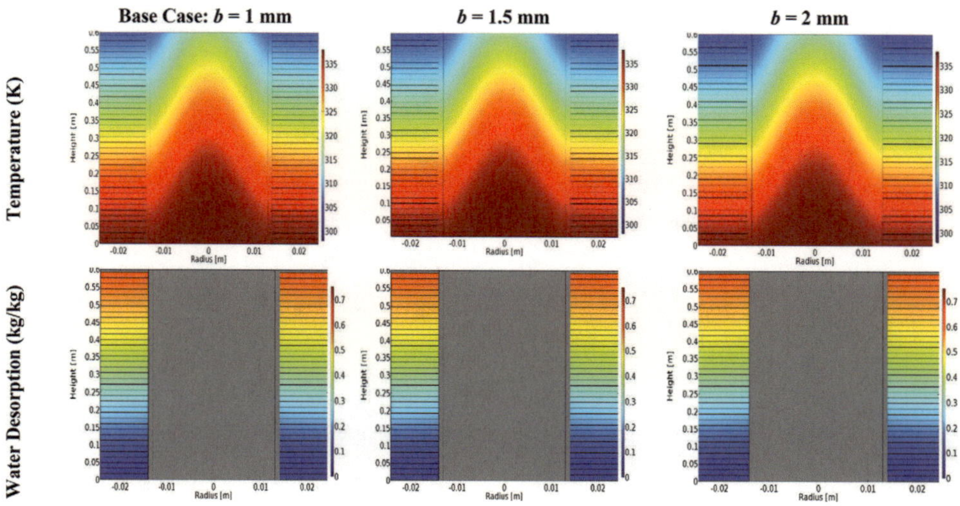

Figure 11.
Effect of tube thickness (b) on desorption 2D temperature and adsorbed water distributions at t = 600 s.

The data in the figures disclose that as the tube thickness increases, neither the bed temperature, pressure or water adsorption shows any significant variance from the baseline case. It is expected that as tube thickness increases, a greater amount of heat should be transferred to the adsorbent bed, and thus should have a positive impact on the overall performance. However, the limited process is controlled by the heat and mass transfer phenomena in the adsorption cooling system. Heat and mass transfer coefficients of the bed are quite small due to low conductivity of adsorbent particles, and high contact resistance between particles and metal tubes/ fins. Therefore, the performance of the adsorbent bed is affected unfavourably by the heat and mass transfer constraints inside the bed, with reduced thermal conductivity of the solid adsorbent, and internal (intra-particle) and external (inter-particle) mass transfer resistances.

4.2.3 Effect of HTF velocity (v)

The effect of HTF velocity on the dependent variables pressure, temperature and amount of water adsorbed within the adsorption cooling system is evaluated for velocity of 0.001 (baseline case), 0.005 and 0.015 m/s. **Figure 12** shows the effects of HTF velocity on simulated transient average bed pressure, average bed temperature and adsorption of water profiles. **Figure 13** presents the simulated temperature and water adsorbed distributions at t = 600 s.

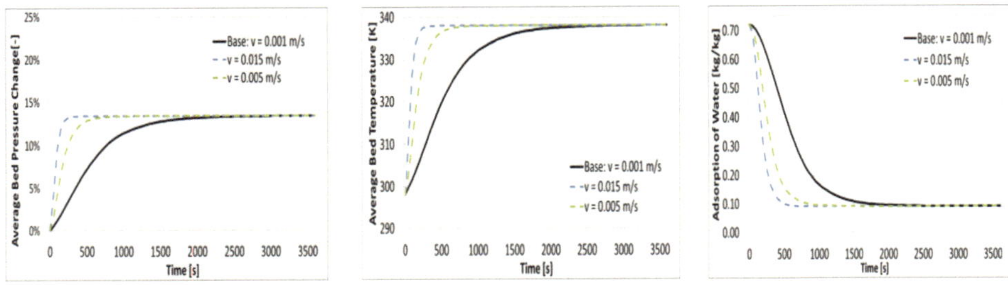

Figure 12.
Effect of HTF velocity (v) on desorption 1D profiles as a function of time for average bed pressure drop, average bed temperature and adsorption of water.

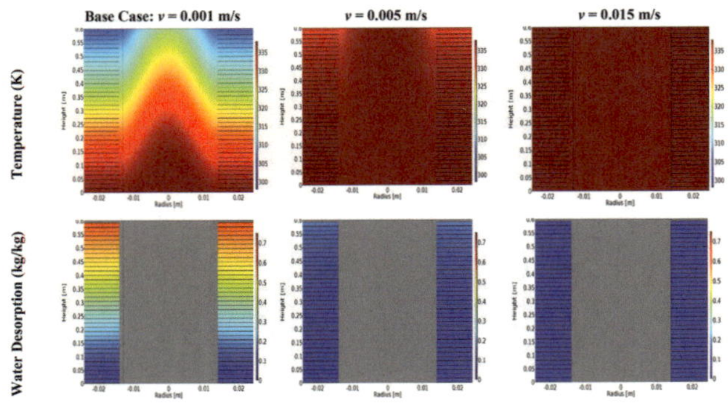

Figure 13.
Effect of HTF velocity on desorption 2D temperature and adsorption of water distributions at t = 600 s.

The results reveal that as the HTF velocity increases, the bed temperature and pressure response increase substantially, as well as the water desorbed. With an increase of HTF velocity from 0.001 to 0.005 m/s the water desorption response almost decreases an upper threshold. Alike, for the fluid flow and thermal properties, as fluid velocity of the tube increases, the temperature profile evolution should rise. The desorption bed produced increased average pressure and temperature values of 1.3 and 1.7% for HTF velocity of 0.055 and 0.015 relative to the baseline case having a HTF velocity of 0.001 m/s, respectively. On the other hand, under the same conditions, the desorption bed produced decreased average adsorbed water values of 18 and 21%, respectively. The adsorbent bed reached a steady state uniform temperature profile at the cycle maximum temperature of 338 K at 500, 700 and 1500 s, respectively. The cycle time increases with the decrease of the HTF velocity in the heat exchanger. An optimal velocity value corresponding to maximum overall performance of the adsorption cooling system, such as a high COP and specific cooling power would be in the range between 0.005 and 0.015 m/s. The HTF velocity should be well selected in order to obtain a good heat transfer efficiency, but also a positive effect on the mass transfer inside the adsorbent bed and on the overall system performance. A too slow HTF will increase cycle time and decrease specific cooling power. A faster velocity will reduce thermal gradient, but will need more pumping energy.

4.2.4 Effect of adsorbent particle diameter (d_p)

The effect of adsorbent particle diameter on the dependent variables pressure, temperature and amount of water adsorbed within the adsorption cooling system is assessed for particle diameter of 2, 3 (baseline case) and 5 mm. **Figure 14** presents

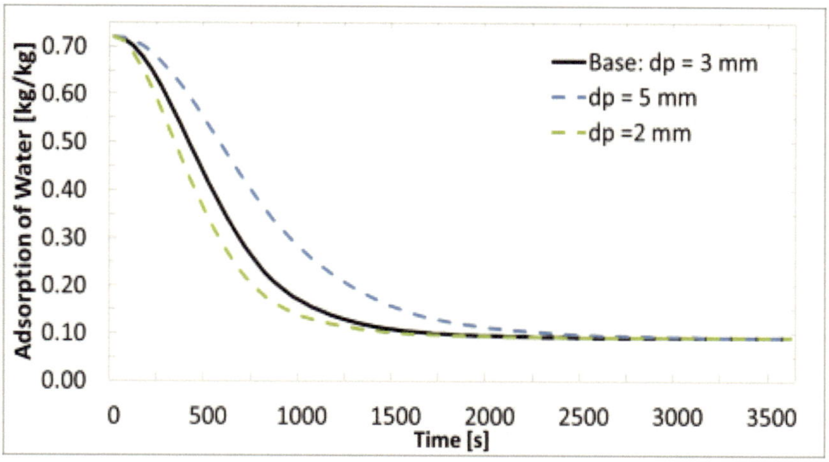

Figure 14.
Effect of particle diameter (d_p) on desorption 1D profiles as a function of time for adsorption of water.

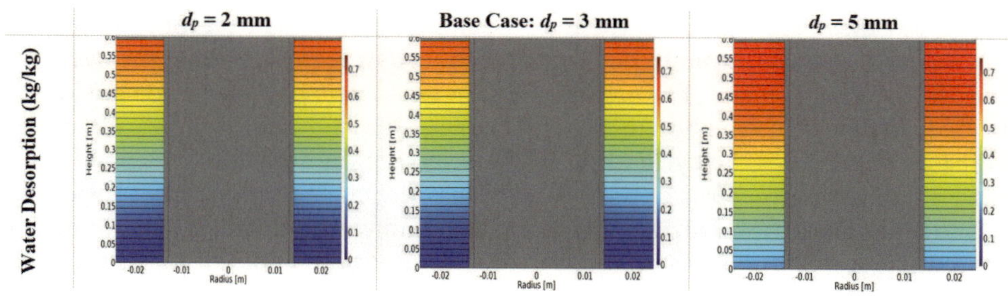

Figure 15.
Effect of particle diameter (d_p) on desorption 2D adsorbed water distributions at t = 600 s.

the effects of particle diameter on simulated transient adsorption of water profile. **Figure 15** shows the water adsorbed distributions at t = 600 s. The temperature and pressure profiles are not shown because of limited space.

The desorption bed produced decreased/increased average temperature and pressure values of about 1.6% for adsorbent particle diameter of 2 mm and 5 mm relative to the baseline case having a particle diameter of 3 mm, respectively. Alternatively, the desorption bed produced decreased average adsorbed water value of 6% for adsorbent particle diameter of 2 mm, and increased value of 22% for adsorbent particle diameter of 5 mm, relative to the baseline case having a particle diameter of 3 mm. The adsorbent bed achieved a steady state uniform temperature profile at the cycle maximum temperature of 338 K at 1000, 1500 and 2000 s, respectively. The smaller the particle diameter the faster the water adsorption response. The kinetics is directly dependent upon the particle packing as seen by the kinetic expressions in Section 2. As the adsorbent particle diameter decreases, the specific area of adsorbent increases, the internal mass transfer resistances decrease. Therefore, the bed temperature, the pressure and the desorption rate increase. The particle diameter should be kept as smaller as possible.

5. Conclusions

In this work, a transient CFD coupled with heat and mass transfer model has been created for a solar adsorption cooling system. Transient simulations have been carried out to investigate the influence of several design and operating parameters during the desorption process. Silica gel and zeolite 13X are both investigated as possible adsorbents, though the study incorporated the working pair silica gel/water given the lower working temperature range required for operation which makes it more suitable for residential cooling applications powered by solar heat. Flow behaviour, heat and mass transfer performances have been analysed in detail. The CFD model was validated against the experimental data using zeolite 13X/water pair available in the literature. Good agreement with experimental results was obtained, which demonstrates the effectiveness of the model. Therefore, the CFD model was utilised in the present study with confidence that it can predict the pressure, temperature and water adsorption of the system accurately.

The key parameters considered in this work were geometrical and operating factors such as bed thickness (l_{bed}), heat exchanger tube thickness (b) and HTF velocity (v), and adsorbent particle diameter (d_p). The results have not only confirmed some observations reported in earlier researches, but also provided an envelope for the optimum design parameters when the operating conditions and geometrical factors were varied for performance enhancement. In order of highest impact on the system, the parameters are listed as follows: $v>d_p>l_{bed}>b$. The results disclosed that from the four parameters; only the first three stated are the most influential factors of performance and significantly change the cycle time. In the design phase, the most important geometric parameter to consider is the bed thickness. As the bed thickness increases, the amount of water adsorbed increases but so does the cycle time. While selecting the adsorbent material, a smaller particle diameter is desired to minimise the cycle time. A lower threshold of 2 mm was identified. During operation, the fluid velocity should be operated at a higher velocity to minimise the cycle time. A upper threshold of 0.005 m/s was identified.

It is concluded that the present modelling approach provides a useful means of identifying significant features, which influence the levels of the pressure drops, temperature and water adsorption from adsorption cooling systems and for assessing the performance characteristic of proposed adsorption chiller configurations. Moreover, the coupled CFD with heat and mass transfer model is a useful tool to simulate and optimise adsorption cooling systems and detect the parameters in the adsorber that are responsible for excessive pressure drops and low performance levels. The effort to alleviate these problems can be directly evaluated. In addition, the influence of any modification that is made to help improve performance characteristics on other operating or geometrical parameters is easily evaluated.

As a final point, this work should lead to accurate design and optimisation of solar/heat-driven adsorption cooling systems in terms of better control of the heat and mass transfer levels over the entire adsorption/desorption operating range and improved overall system performance.

Acknowledgements

Funding for this work was provided by Natural Resources Canada through the Program of Energy Research and Development.

Conflict of interest

The authors declare no conflict of interest.

Nomenclature

a	kinetic constant
b	kinetic constant
b	tube thickness, m
b_t	tube thickness, m
C	specific heat capacity, J/kg K
COP	coefficient of performance
d	fin distance, m
D_o	reference diffusivity, m^2/s
D_m	molecular diffusivity, m^2/s
D-A	Dubinin-Astakhov
d_p	adsorbent particle diameter, m
E_a	activation energy of surface diffusion, J/mol
h_i	convective heat transfer coefficient: tube and fluid, W/m^2 K
h_o	convective heat transfer coefficient: tube and adsorbent, W/m^2 K
ΔH	heat of adsorption, kJ/kg
I	identity tensor
k	thermal conductivity, W/m K
k_m	mass transfer coefficient, 1/s
L	adsorbent bed total length, m
l_{bed}	bed thickness, m
P	pressure, kPa
r_p	particle radius
R	universal gas constant, J/mol K
RD	regular density
R_v	ideal gas constant for water vapour, J/kg K
SPC	specific cooling power, W/kg
T	temperature, K
$\boldsymbol{u}$	velocity vector, m/s
v	fluid velocity, m/s
X	adsorption of water, kg_w/kg_s

Greek letters

Δ	gradient operator
Δ^2	Laplacian operator
ε	void fraction
κ	permeability, m^2
μ	dynamic viscosity, kg/m s
ρ	density, kg/m^3

Subscripts

0	initial condition
a	adsorption
b	adsorbent bed
c	condenser
cold	low temperature setting
d	desorption
e	equilibrium or evaporator

f	fluid
hot	high temperature setting
pl	particle liquid
pv	particle vapour
s	solid adsorbent
v	vapour
t	tube

Author details

Wahiba Yaïci* and Evgueniy Entchev
CanmetENERGY Research Centre, Natural Resources Canada, Ottawa, Ontario, Canada

*Address all correspondence to: wahiba.yaici@canada.ca

References

[1] Saha BB, Akisawa A, Kashiwagi T. Solar/waste heat driven two-stage adsorption chiller: The prototype. Renewable Energy. 2001;**23**:93-101. DOI: 10.1016/S0960-1481(00)00107-5

[2] Tso CY, Chao CYH. Activated carbon, silica-gel and calcium chloride composite adsorbents for energy efficient solar adsorption cooling and dehumidification systems. International Journal of Refrigeration. 2012;**35**: 1626-1638. DOI: 10.1016/j. ijrefrig.2012.05.007

[3] Sah RP, Choudhury B, Das RK. A review on adsorption cooling systems with silica gel and carbon as adsorbents. Renewable and Sustainable Energy Reviews. 2015;**45**:123-134. DOI: 10.1016/j.rser.2015.01.039

[4] Meunier F. Solid sorption: An alternative to CFCs. Heat Recovery CHP. 1993;**13**:289-295. DOI: 10.1016/ 0890-4332(93)90051-V

[5] Meunier F. Solid sorption heat powered cycles for cooling and heat pumping applications. Applied Thermal Engineering. 1998;**18**:715-729. DOI: 10.1016/S1359-4311(97)00122-1

[6] Teng Y, Wang RZ, Wu JY. Study of the fundamentals of adsorption systems. Applied Thermal Engineering. 1997;**17**: 327-338. DOI: 10.1016/S1359-4311(96) 00039-7

[7] Pons M, Meunier F, Cacciola G, Critoph R, Groll M, Puigjaner L, et al. Thermodynamic based comparison of sorption systems for cooling and heat pumping. International Journal of Refrigeration. 1999;**22**:5-17. DOI: S0140-7007(98)00048-6

[8] Hamamoto Y, Alam KC, Saha BB, Koyama S, Akisawa A, Kashiwagi T. Study on adsorption refrigeration cycle utilizing activated carbon fibers. Part 2: Cycle performance evaluation. International Journal of Refrigeration. 2006;**29**:315-327. DOI: 10.1016/j. ijrefrig.2005.06.001

[9] Shmroukh AN, Ali AHH, Ookawara S. Adsorption working pairs for adsorption cooling chillers: A review based on adsorption capacity and environmental impact. Renewable and Sustainable Energy Reviews. 2015;**50**: 445-456. DOI: 10.1016/j. rser.2015.05.035

[10] Kim AS, Lee HS, Moon DS, Kim HJ. Performance control modeling on adsorption desalination using initial time lag (ITL) of individual beds. Desalination. 2016;**369**:1-16. DOI: 10.1016/j.desal.2016.05.004

[11] Zhang LZ, Wang L. Effects of coupled heat and mass transfers in adsorbent on the performance of a waste heat adsorption cooling unit. Applied Thermal Engineering. 1999;**19**: 195-215. DOI: 1359-4311/99

[12] Zhang LZ. A three-dimensional non-equilibrium model for an intermittent adsorption cooling system. Solar Energy. 2000;**69**:27-35. DOI: 0038-092X/00

[13] Solmuş İ, Rees DAS, Yamalı C, Baker D. A two-energy equation model for dynamic heat and mass transfer in an adsorbent bed using silica gel/water pair. International Journal of Heat and Mass Transfer. 2012;**55**:5275-5288. DOI: 10.1016/j.ijrefrig.2011.12.006

[14] Yong L, Sumathy K. Review of mathematical investigation on the closed adsorption heat pump and cooling systems. Renewable and Sustainable Energy Reviews. 2002;**6**:305-338. DOI: 10.1016/S1364-0321(02)00010-2

[15] Teng WS, Leong KC, Chakraborty A. Revisiting adsorption cooling cycle

from mathematical modelling to system development. Renewable and Sustainable Energy Reviews. 2016;**63**: 315-332. DOI: 10.1016/j.rser.2016.05.059

[16] Sakoda A, Suzuki M. Fundamental study on solar powered adsorption cooling system. Journal of Chemical Engineering of Japan. 1984;**17**:52-57. DOI: 10.1252/jcej.17.52

[17] Cho SH, Kim JN. Modeling of a silica gel/water adsorption cooling system. Energy. 1992;**17**:829-839. DOI: 10.1016/0360-5442(92)90101-5

[18] Cacciola G, Restuccia G. Reversible heat pump: A thermodynamic model. International Journal of Refrigeration. 1995;**18**:100-106. DOI: 10.1016/0140-7007(94)00005-I0007/95/0140-7

[19] Guilleminot JJ, Meunier F, Pakleza J. Heat and mass transfer in a non-isothermal fixed bed solid adsorbent external reactor: A uniform pressure–non uniform temperature case. International Journal of Heat and Mass Transfer. 1987;**30**:1595-1606. DOI: 10.1016/0017-9310(87)90304-8

[20] Aittomaki A, Harkonen M. Modelling of zeolite/methanol adsorption heat pump process. Heat Recovery Systems CHP. 1988;**8**:475-482. DOI: 10.1016/0890-4332(88)90053-1

[21] Shelton SV, Wepfer WJ, Miles DJ. External fluid heating of a porous bed. Chemical Engineering Communications. 1988;**71**:39-52. DOI: 10.1080/00986448808940413

[22] Hajji A, Worek WM. Simulation of a regenerative, closed-cycle adsorption cooling/heating system. Energy. 1991;**16**:643-654. DOI: 10.1016/0360-5442(91)90035

[23] Alam KCA, Saha B, Kang YT, Akisawa A, Kashiwagi T. Heat exchanger design effect on the system performance of silica gel adsorption refrigeration systems. International Journal of Heat and Mass Transfer. 2000;**43**:4419-4431. DOI: 10.1016/S0017-9310(00)00072-7

[24] Sun LM, Amar NB, Meunier F. Numerical study on coupled heat and mass transfer in an adsorber with external fluid heating. Heat Recovery Systems & CHP. 1995;**15**:19-29. DOI: 0890-4332(94)E0011-8

[25] Amar NB, Sun LM, Meunier F. Numerical analysis of adsorptive temperature wave regenerative heat pump. Applied Thermal Engineering. 1996;**16**:405-418. DOI: 10.1016/1359-4311(95)00045-3

[26] Solmuş İ, Rees DAS, Yamalı C, Baker D, Kaftanoğlu B. Numerical investigation of coupled heat and mass transfer inside the adsorbent bed of an adsorption cooling unit. International Journal of Refrigeration. 2012;**35**: 652-662. DOI: 10.1016/j.ijrefrig.2011.12.006

[27] Solmuş İ, Yamalı C, Yıldırım C, Bilen K. Transient behavior of a cylindrical adsorbent bed during the adsorption process. Applied Energy. 2015;**142**:115-124. DOI: 10.1016/j.apenergy.2014.12.080

[28] Çağlar A, Yamalı C, Baker DK. Two dimensional transient coupled analysis of a finned tube adsorbent bed for a thermal wave cycle. International Journal of Thermal Sciences. 2013;**73**: 58-68. DOI: 10.1016/j.ijthermalsci.2013.06.009

[29] Çağlar A. The effect of fin design parameters on the heat transfer enhancement in the adsorbent bed of a thermal wave cycle. Applied Thermal Engineering. 2016;**104**:386-393. DOI: 10.1016/j.applthermaleng.2016.05.092

[30] Li XH, Hou XH, Zhang X, Yuan ZX. A review on development of adsorption

cooling—Novel beds and advanced cycles. Energy Conversion and Management. 2015;**94**:221-232. DOI: 10.1016/j.enconman.2015.01.076

[31] COMSOL Inc. COMSOL Multiphysics Software Package Including the CFD, Heat Transfer and Chemical Reaction Engineering Modules. Version 5.2. 2017

[32] Aristov Y, Tokarev M, Freni A, Glaznev I. Kinetics of water adsorption on silica Fuji Davison RD. Microporous and Mesoporous Materials. 2006;**96**: 65-71. DOI: 10.1016/j.micromeso.2006.06.008

[33] Chihara K, Suzuki M. Air drying by pressure swing adsorption. Journal of Chemical Engineering of Japan. 1983;**16**: 293-299. DOI: 10.1252/jcej.16.293

[34] Reid RC, Prausnitz JM, Sherwood TK. Properties of Gases and Liquids. 3rd ed. New York, DC: McGraw-Hill; 1977. ISBN 978-007051790-5

[35] Solmuş İ, Yamalı C, Kaftanoğlu B, Baker D, Çağlar A. Adsorption properties of a natural zeolite–water pair for use in adsorption cooling cycles. Applied Energy. 2010;**87**:2062-2067. DOI: 10.1016/j.apenergy.2009.11.027

[36] Patankar SV. Numerical Heat Transfer and Fluid Flow. 1st ed. New York, DC: Hemisphere Publishing Corporation; 1980. 210 p. DOI: 10.1002/cite.330530323

Chapter 6

Modeling of Fluid-Solid Two-Phase Geophysical Flows

Zhenhua Huang and Cheng-Hsien Lee

Abstract

Fluid-solid two-phase flows are frequently encountered in geophysical flow problems such as sediment transport and submarine landslides. It is still a challenge to the current experiment techniques to provide information such as detailed flow and pressure fields of each phase, which however is easily obtainable through numerical simulations using fluid-solid two-phase flow models. This chapter focuses on the Eulerian-Eulerian approach to two-phase geophysical flows. Brief derivations of the governing equations and some closure models are provided, and the numerical implementation in the finite-volume framework of OpenFOAM® is described. Two applications in sediment transport and submarine landslides are also included at the end of the chapter.

Keywords: granular flows, submarine landslides, sediment transport, scour, continuum model, OpenFOAM®

1. Introduction

Fluid-solid two-phase flows are important in many geophysical problems such as sediment erosion, transport and deposition in rivers or coastal environment, debris flows, scour at river or marine structures, and submarine landslides. Behaviors of fluid-solid two-phase flows are very different from those of liquid-gas two-phase flows where bubbles are dispersed in the liquid or droplets dispersed in the gas. Vast numbers of experiments on various scales have been carried out for different applications of fluid-solid two-phase flows; these experiments have advanced our understanding of bulk behaviors of some important flow characteristics. However, development of measurement techniques suitable for collecting data that contribute to understanding important physics involved in fluid-solid two-phase flows is a still-evolving science. With the modern computer technology, many data that are not obtainable currently in the experiment can be easily produced by performing time-dependent, multidimensional numerical simulations. Of course, empirical closure models required to close the governing equations still need high-quality experimental data for model validation.

Numerical approaches to two-phase flows include Eulerian-Eulerian approach, direct numerical simulations (DNS) based on Eulerian-Lagrangian formulations (Lagrangian point-particle approach), and fully resolved DNS approach [1]. Fully resolved DNS can resolve all important scales of the fluid and particles, but these simulations are currently limited to about 10 k uniform-size spheres on a Cray XE6 with 2048 cores [2], and it is not practical to use this method to model large-scale geophysical flow problems in the foreseeable future [1]. Lagrangian point-particle approach uses Eulerian formulation for the fluid phase and Lagrangian formulation for tracking the instantaneous positions of the particles. Lagrangian point-particle simulations make use of semiempirical relationships to provide both hydrodynamic force and torque acting on each particle and thus avoid modeling processes on scales smaller than Kolmogorov scale [1], making it possible to include more particles and run in a domain larger than that for fully resolved DNS. The application of Lagrangian point-particle approach is crucially dependent on the availability and accuracy of such semiempirical relationships. A recent study shows that good results can be obtained for about 100k uniform-size spherical particles in a vertical channel flow [3]; however, using this approach to investigate large-scale two-phase flow problems is still beyond the current computing capacity. Two-phase Eulerian-Eulerian approach treats both the fluid and particle phases as continuum media and is suitable for solving large-scale two-phase flow problems.

Eulerian-Eulerian two-phase flow models based on large-eddy-simulations solve a separate set of equations describing conservation of mass, momentum, and kinetic energy for each phase [4–7] and thus have the potential to consider all important processes involved in the interactions between the two phases through parameterization of particle-scale processes. This chapter introduces the basics of Eulerian-Eulerian two-phase flow modeling, its implementation in the finite-volume framework of OpenFOAM®, and two applications in geophysical flow problems.

2. Governing equations for fluid-solid two-phase flows

Let us consider a mixture of fluid and solid particles. Fluid can be gas, water, or a mixture of water and gas. In DNS and Lagrangian point-particle approaches to two-phase flows, the flow field is solved by solving the Navier-Stokes equations, and the motion of each particle is determined by the Newton's equation of motion. In Eulerian-Eulerian two-phase flow approaches, however, the motions of individual particles are not of the interest, and the focus is on the macroscopic motion of the fluid and solid particles instead. For this purpose, the solid particles are modeled as a continuum mass through an ensemble averaging operation, which is based on the existence of possible equivalent realizations. After taking ensemble average, the mixture of fluid and particles consists of two continuous phases: the fluid (water, gas, or a mixture of water and gas) is the fluid phase, and the solid particle is the solid phase. Both phases are incompressible. The motions of the fluid and solid phases are governed by their own equations, which are obtained by taking ensemble average of the microscopic governing equations for each phase [8]. Even though some aspects of fluid-solid interaction can be considered through the ensemble average, the ensemble averaging operation itself, however, does not explicitly introduce any turbulent dispersion in the resulting equations. To consider the turbulent dispersion in the Eulerian-Eulerian description of the fluid-solid two-phase flows, another averaging operation (usually a Favre average) is needed to consider the correlations of turbulent components [5, 9].

2.1 Ensemble averaged equations

At the microscopic scale, the fluid-solid mixture is a discrete system. The purpose of performing an ensemble averaging operation is to derive a set of equations describing this discrete system as a continuous system at the macroscopic scale, where the typical length scale should be much larger than one particle diameter.

In the Eulerian-Eulerian approach to two-phase flows, it is assumed that the equations governing the motion of phase k (for the fluid phase $k = f$ and for the solid phase $k = s$) at the microscopic scale are the following equations for the conservation of mass and momentum [8, 10]:

$$\frac{\partial \rho_k}{\partial t} + \nabla \cdot \mathbf{u}_k = 0, \tag{1}$$

and

$$\frac{\partial \rho \mathbf{u}_k}{\partial t} + \nabla \cdot (\rho_k \mathbf{u}_k \mathbf{u}_k) = \nabla \cdot \mathbf{T}_k + \rho \mathbf{g}, \tag{2}$$

where ρ_k is the density, $\mathbf{u}_k$ is the velocity, and $\mathbf{g}$ is the acceleration due to gravity. The stress tensor $\mathbf{T}_k$ includes two components:

$$\mathbf{T}_k = -p_k \mathbf{I} + \tau_k \tag{3}$$

where p^k is the microscopic pressure and τ^k is the microscopic stress tensor.

Because the fluid phase and the solid phase are immiscible, at any time t, a point in space $\mathbf{x}$ can be occupied only by one phase, not both. This fact can be described mathematically by the following phase function $c_k(\mathbf{x}, t)$ for phase k:

$$c_k(\mathbf{x}, t) = \begin{cases} 1, & \text{if the point } \mathbf{x} \text{ is occupied by phase } k \\ 0, & \text{if the point } \mathbf{x} \text{ is not occupied by phase } k \end{cases}. \tag{4}$$

The volumetric concentration of phase k is directly related to the probability of occurrence of phase k at a given location $\mathbf{x}$ at the time t and can be obtained by taking ensemble average of c_k. Using the phase function given in Eq. (4), the volumetric concentration of phase k is obtained by taking the ensemble average of c_k, denoted by $\langle c_k \rangle$. The operator $\langle \cdots \rangle$ means taking an ensemble average of its argument.

There are several methods to derive the ensemble averaged equations governing the motion of phase k. This chapter treats the phase function as a general function and uses it to define the derivatives of the phase function c_k with respect to time and space and the equation governing the evolution of c_k. As stated in Drew [8], the phase function c_k can be treated as a generalized function whose derivative can be defined in terms of a set of test functions. These test functions must be sufficiently smooth and have compact support so that the integration of a derivative of the phase function, weighed with the test function, is finite. The equation describing the evolution of c_k is

$$\frac{\partial c_k}{\partial t} + \mathbf{u}^i \cdot \nabla c_k = 0, \tag{5}$$

where $\mathbf{u}^i$ is the velocity of the interface between the region occupied by the fluid phase and the region occupied by the solid phase. It is stressed here that ∇c_k is zero except at the interface between two phases where ∇c_k behaves like a delta-function [8].

The ensemble averaged equations governing the motion of phase k are obtained by multiplying Eqs. (1) and (2) with c_k and performing an ensemble average operation on every term in the resulting equations. When performing ensemble average operations, Reynolds' rules for algebraic operations, Leibniz' rule for time derivatives, Gauss' rule for spatial derivatives, and the following two identities are used:

$$c_k \frac{\partial \phi_k}{\partial t} = \frac{\partial c_k \phi_k}{\partial t} - \phi_k \frac{\partial c_k}{\partial t} = \frac{\partial c_k \phi_k}{\partial t} + \phi_k \mathbf{u}^i \cdot \nabla c_k, \tag{6}$$

and

$$c_k \nabla \cdot (\rho_k \mathbf{u}_k) = \nabla \cdot (c_k \rho_k \mathbf{u}_k) - (\rho_k \mathbf{u}_k) \cdot \nabla c_k. \tag{7}$$

The resulting equations governing the ensemble average motion of phase k are [8]

$$\frac{\partial \langle c_k \rho_k \rangle}{\partial t} + \nabla \cdot \langle c_k \rho_k \mathbf{u}_k \rangle = \left\langle \rho_k \left(\mathbf{u}_k - \mathbf{u}^i \right) \cdot \nabla c_k \right\rangle, \tag{8}$$

and

$$\frac{\partial \langle c_k \rho_k \mathbf{u}_k \rangle}{\partial t} + \nabla \cdot \langle c_k \rho_k \mathbf{u}_k \mathbf{u}_k \rangle = \nabla \cdot \langle c_k \mathbf{T}_k \rangle + \langle c_k \rho_k \mathbf{g} \rangle + \tilde{\mathbf{m}}_k \tag{9}$$

with

$$\tilde{\mathbf{m}}_k = \left\langle \rho_k \mathbf{u}_k \left(\mathbf{u}_k - \mathbf{u}^i \right) - \mathbf{T}_k \cdot \nabla c_k \right\rangle, \tag{10}$$

Note that ∇c_k is not zero only on the interface of the region occupied by phase k (grain boundary). For the fluid-solid two-phase flows, the interface of phase k must satisfy the no-slip and no-flux conditions; therefore, $\mathbf{u}_k - \mathbf{u}^i = 0$. As a result, the right-hand side of Eq. (8) is zero and

$$\tilde{\mathbf{m}}_k = -\langle \mathbf{T}_k \cdot \nabla c_k \rangle, \tag{11}$$

which is the density of the interfacial force [8]. Physically, $\mathbf{T}^k \cdot \nabla c_k$ is the microscopic density of the force acting on a surface whose normal direction is defined by ∇c_k.

After using Eq. (3) for $\mathbf{T}_k$ in Eq. (9), the ensemble averaged equations can be further written in terms of the ensemble averaged qualities describing the motion of phase k as

$$\frac{\partial \tilde{c} \tilde{\rho}_k}{\partial t} + \nabla \cdot \left[\tilde{c} \tilde{\rho}_k \hat{\mathbf{u}}_k \right] = 0 \tag{12}$$

and

$$\frac{\partial \tilde{c}_k \tilde{\rho}_k \hat{\mathbf{u}}_k}{\partial t} + \nabla \cdot \left[\tilde{c} \tilde{\rho}_k \hat{\mathbf{u}}_k \hat{\mathbf{u}}_k \right] = \tilde{c} \tilde{\rho}_k \mathbf{g} + \nabla \cdot \left[\tilde{c} \left(\tilde{p}_k \mathbf{I} + \tilde{\tau}_k \right) \right] + \nabla \cdot \left[\tilde{c} \tilde{\tau}'_k \right] + \tilde{\mathbf{m}}_k, \tag{13}$$

where $\tilde{c}_k = \langle c_k \rangle$ is the volumetric concentration of phase k. Other ensemble averaged quantities used in Eqs. (12) and (13) to describe the motion of phase k at the macroscopic scale are density $\tilde{\rho}_k$, pressure $\tilde{p}_k$, stress tensor $\tilde{\tau}_k$, and velocity $\hat{\mathbf{u}}_k$, defined by

$$\tilde{\rho}_k = \frac{\langle c_k \rho_k \rangle}{\langle c_k \rangle}, \quad \tilde{p}_k = \frac{\langle c_k p_k \rangle}{\langle c_k \rangle}, \quad \tilde{\tau}_k = \frac{\langle c_k \tau_k \rangle}{\langle c_k \rangle}, \quad \hat{\mathbf{u}}_k = \frac{\langle c_k \rho_k \mathbf{u}_k \rangle}{\langle c_k \rho_k \rangle} \tag{14}$$

and $\tilde{\boldsymbol{t}}'_k$ represents the c-weighted ensemble average of microscopic momentum flux associated with the fluctuation of the velocity $\mathbf{u}_k$ around the ensemble averaged velocity $\hat{\mathbf{u}}_k$

$$\tilde{\tau}'_k = -\frac{\langle c_k \rho_k \mathbf{u}'_k \mathbf{u}'_k \rangle}{\langle c_k \rangle}, \quad \mathbf{u}'_k = \mathbf{u}_k - \hat{\mathbf{u}}_k \tag{15}$$

For compressible materials $\tilde{\rho}_k$ is not a constant. However, for incompressible materials

$$\tilde{\rho}_k = \frac{\langle c_k \rho_k \rangle}{\langle c_k \rangle} \equiv \rho_k, \quad \hat{\mathbf{u}}_k = \frac{\langle c_k \rho_k \mathbf{u}_k \rangle}{\langle c_k \rho_k \rangle} = \frac{\langle c_k \mathbf{u}_k \rangle}{\langle c_k \rangle} \tag{16}$$

Now we examine the limiting case where the fluid-solid system is at its static state. Because the phase functions for the two phases satisfy $c_f + c_s = 1$, both phases are not moving, and $\tilde{\mathbf{m}}_f + \tilde{\mathbf{m}}_s = 0$, the governing equations reduce to

$$0 = \left(1 - \tilde{c}_s\right)\tilde{\rho}_f \mathbf{g} - \nabla\left[\left(1 - \tilde{c}_s\right)\tilde{p}_f\right] - \tilde{\mathbf{m}}_s, \tag{17}$$

for the fluid phase, and

$$0 = \tilde{c}_s \tilde{\rho}_s \mathbf{g} - \nabla\left[\tilde{c}_s \tilde{p}_s\right] + \tilde{\mathbf{m}}_s, \tag{18}$$

for the solid phase.

Because $\tilde{p}_f$ is the hydrostatic pressure in this case, i.e., $\nabla \tilde{p}_f = \tilde{\rho}_f \mathbf{g}$, it then follows that

$$\tilde{\mathbf{m}}_s = \tilde{p}_f \nabla \tilde{c}_s \tag{19}$$

which, physically, is the buoyancy acting on the solid phase. Now Eq. (18) becomes

$$0 = \tilde{c}_s \tilde{\rho}_s \mathbf{g} - \nabla\left[\tilde{c}_s \tilde{p}_s\right] + \tilde{p}_f \nabla \tilde{c}_s \tag{20}$$

which states that the weight of the solid particles is supported by the buoyancy and the interparticle forces. Therefore, the ensemble pressure of the solid phase can be written as $\tilde{p}_s = \tilde{p}_f + \breve{p}_s$, with $\tilde{p}_f$ being the total fluid pressure and $\breve{p}_s$ accounting for the contributions from other factors such as collision and enduring contact to the ensemble averaged pressure.

For brevity of the presentation, we shall denote simply c_s by c as well c_f by $1 - c$ and drop the symbols representing the ensemble averages hereinafter. The ensemble averaged equations governing the motion of the fluid phase are

$$\frac{\partial (1-c)\rho_f}{\partial t} + \nabla \cdot \left[(1-c)\rho_f \mathbf{u}_f\right] = 0, \tag{21}$$

and

$$\begin{aligned} \frac{\partial (1-c)\rho_f \mathbf{u}_f}{\partial t} &+ \nabla \cdot \left[(1-c)\rho_f \mathbf{u}_f \mathbf{u}_f\right] \\ &= (1-c)\rho_f \mathbf{g} + \nabla \cdot \left[(1-c)\left(-p_f \mathbf{I} + \tau_f\right)\right] + \nabla \cdot \left[(1-c)\tau'_f\right] - \mathbf{m}. \end{aligned} \tag{22}$$

The ensemble averaged equations governing the motion of the solid phase are

$$\frac{\partial c\rho_s}{\partial t} + \nabla \cdot (c\rho_s \mathbf{u}_s) = 0, \tag{23}$$

and

$$\frac{\partial c\rho_s \mathbf{u}_s}{\partial t} + \nabla \cdot \left[c\rho_f \mathbf{u}_s \mathbf{u}_s\right] = \rho_s c\mathbf{g} + \nabla \cdot \left[c\left(-p_f \mathbf{I} - p_s \mathbf{I} + \tau_s\right)\right] + \nabla \cdot \left(c\tau_s'\right) + \mathbf{m}. \tag{24}$$

where p_s denotes the contributions from interparticle interactions such as collision and enduring contact to the ensemble averaged pressure of the solid phase.

To close the equations for the fluid and solid phases, closure models are needed for τ_s', τ_f', τ_s, τ_f, p_s, and $\mathbf{m}$.

It is remarked here that the definitions of the ensemble averages given in Eq. (14) do not consider the contribution from the correlations between the fluctuations of the velocities and the fluctuations of phase functions at microscopic scale; therefore, the effects of turbulent dispersion are not directly included in the ensemble averaged equations describing the motion of the each phase. In the literature, two approaches have been used to consider the turbulent dispersion: (i) considering the correlation between the fluctuations of $\langle c_k \rangle$ and $\mathbf{u}_f$ associated with the turbulent flow [9] and (ii) including a term in the model for $\mathbf{m}$ to account for the turbulent dispersion [8]. This chapter considers the turbulent dispersion using the first approach in the next section by taking another Favre averaging operation.

In the absence of the turbulent dispersion from $\mathbf{m}$, the interphase force $\mathbf{m}$ should include the so-called general buoyancy $p_f \nabla c$ and a component $\mathbf{f}$ which includes drag force, inertial force, and lift force

$$\mathbf{m} = \mathbf{f} + p_f \nabla c \equiv \mathbf{f} - c\nabla p_f + \nabla\left(cp_f\right). \tag{25}$$

This expression for $\mathbf{m}$ has been derived by [11] using a control volume/surface approach. For most fluid-solid two-phase geophysical flows, the drag force dominates $\mathbf{f}$ [9] and thus $\mathbf{f}$ can be modeled by

$$\mathbf{f} = c\rho_s \frac{\mathbf{u}_f - \mathbf{u}_s}{\tau_p}, \tag{26}$$

where τ_p is the so-called particle response time (i.e., a relaxation time of the particle to respond the surrounding flow). As expected, the particle response time should be related to drag coefficient and grain Reynolds number.

2.2 Favre averaged equations

The volumetric concentration and the velocities can be written as

$$c = \bar{c} + c'', \ p_f = \bar{p}_f + p_f'', \ \mathbf{u}_f = \bar{\mathbf{u}}_f + \mathbf{u}_f'', \ \mathbf{u}_s = \bar{\mathbf{u}}_s + \mathbf{u}_s'', \tag{27}$$

where the Favre averages are defined as

$$\bar{\rho}_s = \frac{\overline{c\rho_s}}{\bar{c}}, \quad \bar{\rho}_f = \frac{\overline{(1-c)\rho_f}}{1-\bar{c}}, \ \bar{\mathbf{u}}_s = \frac{\overline{c\rho_s \mathbf{u}_s}}{\overline{c\rho_s}}, \ \bar{\mathbf{u}}_f = \frac{\overline{(1-c)\rho_f \mathbf{u}_f}}{\overline{(1-c)\rho_f}}, \tag{28}$$

and the overline stands for an integration with respect to time over a time scale longer than small-scale turbulent fluctuations but shorter than the variation of the mean flow field.

The averaged equations for the mean flow fields of the two phases are obtained by taking the following steps: (i) substituting Eq. (25) with Eq. (26) in Eqs. (22) and (24), (ii) substituting Eq. (27) in the equations obtained at step (i), and (iii) taking average of the equations obtained at step (ii) to obtain the following equations:

$$\frac{\partial \overline{\rho}_f(1-\overline{c})}{\partial t} + \nabla \cdot \left[\overline{\rho}_f(1-\overline{c})\overline{\mathbf{u}}_f\right] = 0, \tag{29}$$

$$\begin{aligned}\frac{\partial \overline{\rho}_f(1-\overline{c})\overline{\mathbf{u}}_f}{\partial t} &+\nabla \cdot \left[\overline{\rho}_f(1-\overline{c})\overline{\mathbf{u}}_f\overline{\mathbf{u}}_f\right] = \overline{\rho}_f(1-\overline{c})\mathbf{g} - (1-\overline{c})\nabla\overline{p}_f + \overline{c''\nabla p_f''} \\ &+\nabla \cdot \overline{(1-c)\left(\tau_f + \tau_f' + \tau_f''\right)} - \overline{c}\overline{\rho}_s\frac{\overline{\mathbf{u}}_f - \overline{\mathbf{u}}_s}{\tau_p} - \frac{\overline{\rho}_s}{\tau_p}\left(\overline{c\mathbf{u}_f''}\right),\end{aligned} \tag{30}$$

for the fluid phase, with τ_f'' being defined by

$$\tau_f'' = -\rho_f \mathbf{u}_f''\mathbf{u}_f'', \tag{31}$$

and

$$\frac{\partial \overline{\rho}_s\overline{c}}{\partial t} + \nabla \cdot \left[\overline{\rho}_s\overline{c}\overline{\mathbf{u}}_s\right] = 0, \tag{32}$$

$$\begin{aligned}\frac{\partial \overline{\rho}_s\overline{c}\overline{\mathbf{u}}_s}{\partial t} &+\nabla \cdot \left[\overline{\rho}_s\overline{c}\overline{\mathbf{u}}_s\overline{\mathbf{u}}_s\right] = \overline{\rho}_s\overline{c}\mathbf{g} - \overline{c}\nabla\overline{p}_f - \overline{c''\nabla p_f''} - \nabla\overline{cp_s} \\ &+\nabla \cdot \overline{c\left(\tau_s + \tau_s' + \tau_s''\right)} + \overline{c}\overline{\rho}_s\frac{\overline{\mathbf{u}}_f - \overline{\mathbf{u}}_s}{\tau_p} + \frac{\overline{\rho}_s}{\tau_p}\left(\overline{c\mathbf{u}_f''}\right),\end{aligned} \tag{33}$$

for the solid phase, with τ_s'' being defined by

$$\tau_s'' = -\rho_s \mathbf{u}_s''\mathbf{u}_s'' \tag{34}$$

It is remarked here that the terms $(1-\tilde{c})\nabla\tilde{p}_f$ in Eq. (30) and $\tilde{c}\nabla\tilde{p}_f$ in Eq. (33) have been obtained by using the expression for $\mathbf{m}$ given in Eq. (25).

In order to close these averaged equations, closure models are required for the following terms: $\overline{c\left(\tau_s + \tau_s' + \tau_s''\right)}$, $\overline{(1-c)\left(\tau_f + \tau_f' + \tau_f''\right)}$, $\overline{c\mathbf{u}_f''}$, and $\overline{c''\nabla p_f''}$. The last term can be neglected based on an analysis of their orders of magnitude by Drew [12]. The term $\overline{c\mathbf{u}_f''}$ is approximated by the following gradient transport hypotheses:

$$\overline{c\mathbf{u}_f''} = -\frac{\nu_{ft}}{\sigma_c}\nabla c \tag{35}$$

where ν_{ft} is the eddy viscosity and σ_c is the Schmidt number, which represents the ratio of the eddy viscosity of the fluid phase to the eddy diffusivity of the solid phase. Furthermore, the following approximations are introduced:

$$\overline{c\left(\tau_s + \tau_s' + \tau_s''\right)} = \overline{c}\overline{\tau}_s, \quad \overline{(1-c)\left(\tau_f + \tau_f' + \tau_f''\right)} = (1-\overline{c})\overline{\tau}_f, \quad \overline{cp_s} = \overline{c}\overline{p}_s \tag{36}$$

For brevity of the presentation, the symbols representing Favre averages are dropped hereinafter, and the final equations governing the conservation of mass and momentum of each phase are

$$\frac{\partial \rho_f(1-c)}{\partial t} + \nabla \cdot \rho_f(1-c)\mathbf{u}_f = 0, \tag{37}$$

$$\begin{aligned}\frac{\partial\rho_f(1-c)\mathbf{u}_f}{\partial t} &+\nabla\cdot\left[\rho_f(1-c)\mathbf{u}_f\mathbf{u}_f\right]\\ &=\rho_f(1-c)\mathbf{g}-(1-c)\nabla p_f+\nabla\cdot(1-c)\tau_f\\ &-\left\{c\rho_s\frac{\mathbf{u}_f-\mathbf{u}_s}{\tau_p}+\frac{\rho_s}{\tau_p}\frac{\nu_{ft}}{\sigma_c}\nabla c\right\},\end{aligned} \tag{38}$$

for the fluid phase and

$$\frac{\partial\rho_s c}{\partial t}+\nabla\cdot\rho_s c\mathbf{u}_s=0, \tag{39}$$

$$\begin{aligned}\frac{\partial\rho_s c\mathbf{u}_s}{\partial t} &+\nabla\cdot\left(\rho_s c\mathbf{u}_s\mathbf{u}_s\right)\\ &=\rho_s c g-c\nabla p_f-\nabla\left(cp_s\right)+\nabla\cdot c\tau_s\\ &+\left\{c\rho_s\frac{\left(\mathbf{u}_f-\mathbf{u}_s\right)}{\tau_p}-\frac{\rho_s}{\tau_p}\frac{\nu_{ft}}{\sigma_c}\nabla c\right\},\end{aligned} \tag{40}$$

for the solid phase.

3. Closure models

3.1 Stresses for the fluid phase

The stress tensor for the fluid phase τ_f includes two parts: a part for the viscous stress, τ_f^v, and the other part for the turbulent Reynolds stress, τ_f^t

$$\tau_f=\tau_f^v+\tau_f^t \tag{41}$$

The viscous stress tensor τ_f^v is usually computed by

$$\tau_f^v=-\rho_f\left(\frac{2}{3}\nu_f\nabla\cdot\mathbf{u}_f\right)\mathbf{I}+2\rho_f\nu_f\mathbf{D}_f \tag{42}$$

where ν_f is the kinematic viscosity of the fluid phase and $\mathbf{D}_f=\left[\nabla\mathbf{u}_f+\left(\nabla\mathbf{u}_f\right)^T\right]/2$, where the superscript T denotes a transpose. Some studies [13] suggested modifying ν_f to consider the effect of the solid phase; other studies [14], however, obtained satisfactory results even without considering this effect.

The stress tensor τ_f^t is related to the turbulent characteristics, which need to be provided by solving a turbulent closure model such as $k-\epsilon$ or $k-\omega$ model. For a $k-\epsilon$ model with low-Reynolds-number correction [15], τ_f^t can be computed by

$$\tau_f^t=-\rho_f\left(\frac{2}{3}k+\frac{2}{3}\nu_f^t\nabla\cdot\mathbf{u}_f\right)\mathbf{I}+2\rho_f\nu_f^t\mathbf{D}_f \tag{43}$$

where k is the turbulence kinetic energy and ν_f^t is the eddy viscosity of the fluid phase, given by

$$\nu_f^t=f_\mu C_\mu k^2/\epsilon \tag{44}$$

with ϵ being the turbulent dissipation of the fluid phase to be provide by solving the $k-\epsilon$ equation. The coefficient $f_\mu = \exp\left[-3.4/(1+\mathrm{Re}_t/50)^2\right]$ represents the low-Reynolds-number correction with $\mathrm{Re}_t = k^2/\nu_f\epsilon$. The coefficient C_μ is usually assumed to be a constant.

The equations governing k and ϵ are similar to those for clear water [15]

$$\frac{\partial\rho_f(1-c)k}{\partial t} + \nabla\cdot\left[\rho_f(1-c)\mathbf{u}_f k\right] = (1-c)\boldsymbol{t}_f : \nabla\mathbf{u}_f - \rho_f(1-c)\epsilon + \nabla\cdot\left[\rho_f\frac{\nu_f^t}{\sigma_c}(1-c)k\right] - \left\{\left(\rho_s-\rho_f\right)\frac{\nu_f^t}{\sigma_c}\nabla c\cdot\mathbf{g} + \frac{2\rho_s c(1-\alpha)k}{\tau_p}\right\}, \tag{45}$$

and

$$\frac{\partial\rho_f(1-c)\epsilon}{\partial t} + \nabla\left[\rho_f(1-c)\mathbf{u}_f\epsilon\right] = \frac{\epsilon}{k}\left[C_{\epsilon1}f_1(1-c)\tau_f : \nabla\mathbf{u}_f - C_{\epsilon2}f_2\rho_f(1-c)\epsilon\right] + \nabla\cdot\left[\rho_f\frac{\nu_f^t}{\sigma_\epsilon}(1-c)\epsilon\right] - \frac{\epsilon}{k}C_{\epsilon3}\left\{\left(\rho_s-\rho_f\right)\frac{\nu_f^f}{\sigma_c}\nabla c\cdot\mathbf{g} + \frac{2\rho_s c(1-\alpha)k}{\tau_p}\right\}, \tag{46}$$

where coefficients $C_{\epsilon1}$, $C_{\epsilon2}$, σ_ϵ, σ_k, and f_2 are model parameters, whose values can be taken the same as those in the $k-\epsilon$ model for clear fluid under low-Reynolds-number conditions [15]. There are two terms inside the curly brackets, and both terms account for the turbulence modulation by the presence of particles: the first term is associated with the general buoyancy, and the second term is due to the correlation of the fluctuating velocities of solid and fluid phases. $C_{\epsilon3} = 1$ is usually adopted in the literature [28]; however, it is remarked that the value of $C_{\epsilon3}$ is not well understood at the present and a sensitivity test to understand how the value of this $C_{\epsilon3}$ on the simulation results is recommended. The parameter α reflects the correlation between the solid-phase and fluid-phase turbulent motions and is given by

$$\alpha = \left(1 + \frac{\tau_p}{\min(\tau_l, \tau_c)}\right)^{-1}, \tag{47}$$

where $\tau_l = 0.165k/\epsilon$ is a time scale for the turbulent flow and τ_c is a time scale for particle collisions given by [16]

$$\tau_c = \left[\left(\frac{c_{\mathrm{rcp}}}{c}\right)^{\frac{1}{3}} - 1\right] d\left(\frac{\rho_s}{p_s}\right)^{1/2} \tag{48}$$

with c_{rcp} being the random close packing fraction and d being the particle diameter. c_{rcp} is 0.634 for spheres [17]. The term $\left(c_{\mathrm{rcp}}/c\right)^{1/3} - 1$ is related to the ratio of the mean free dispersion distance to the diameter of the solid particle.

It is remarked here that the presence of solid particles in the turbulent flow may either enhance (for large particles) or reduce (for small particles) the turbulence [18]. The $k-\epsilon$ model given here can only reflect the reduction of turbulence and thus is not suitable for problems with large particles. Other turbulence models [7, 18] include a term describing the enhancement of turbulence; however, including that term in the present model may induce numerical instability in some cases.

3.2 Stresses for the solid phase

The closure models for p_s and τ_s used in Lee et al. [16] will be described here. In order to cover flow regimes with different solid-phase concentrations (dilute flows, dense flows, and compact beds), Lee et al. [16] suggested the following model for p_s:

$$p_s = p_s^t + p_s^r + p_s^e, \tag{49}$$

where p_s^t accounts for the turbulent motion of solid particles (important for dilute flows); p_s^r reflects the rheological characteristics of dense flows and includes the effects such as fluid viscosity, enduring contact, and particle inertial; p_s^e accounts for the elastic effect, which is important when the particles are in their static state in a compact bed.

For solid particles in a compact bed, the formula proposed by Hsu et al. [19] can be used to compute p_s^e

$$p_s^e = K[\max(c - c_o, 0)]^\chi \left\{1 + \sin\left[\max\left(\frac{c - c_o}{c_{\mathrm{rcp}} - c_o}, 0\right)\pi - \frac{\pi}{2}\right]\right\}, \tag{50}$$

where c_o is random loose packing fraction and coefficients K and χ are model parameters. For spheres, c_o ranges from 0.54 to 0.634, depending on the friction [17]. The coefficient K is associated with the Young's modulus of the compact bed, and the other terms are related to material deformation.

The closure models for p_s^r and p_s^t are closely related to the stress tensor and the visco-plastic rheological characteristics for the solid phase. The stress tensor for the solid phase can be computed by

$$\boldsymbol{t}_s = -\left(\frac{2}{3}\rho_s\nu_s\nabla\cdot\mathbf{u}_s\right) + 2\rho_s\nu_s\mathbf{D}_s, \tag{51}$$

The kinematic viscosity of the solid phase ν_s is computed by the sum of two terms:

$$\nu_s = \nu_s^v + \nu_s^t, \tag{52}$$

where ν_s^v and ν_s^t represent the visco-plastic and turbulence effects, respectively. This model for ν_s can consider both the turbulence behavior (for dilute flows) and the visco-plastic behavior (for dense flows and compact beds).

Based on an analysis of heavy and small particles in homogeneous steady turbulent flows, Hinze [20] suggests that p_s^t and ν_s^t can be computed by

$$p_s^t = \frac{2}{3}\rho_s\alpha k, \tag{53}$$

and

$$\nu_s^t = \alpha\nu_f^t. \tag{54}$$

where the coefficient α is the same as that in Eqs.(45) and (46).

For dense fluid-solid two-phase flows, the visco-plastic rheological characteristics depend on a dimensionless parameter $I = I_v + aI_i^2$, where I_v is the viscous number, I_i is the inertial number, and a is a constant [21]. The viscous number is defined by $I_v = 2\rho_f\nu_f D^s/cp_s$ where ν_f is the kinematic viscosity of the fluid and D^s is the second invariant of the strain rate. Physically, the viscous number describes the ratio of the viscous stress to the quasi-static shear stress associated with the weight (resulting from the enduring contact). The inertial number is defined by $I_i = 2dD^s/\sqrt{cp_s/\rho_s}$, which describes the ratio of the inertial stress to the quasi-static stress. The relative importance of the inertial number to the viscous number can be measured by the Stokes number $\mathrm{st}_v = I_i^2/I_\nu$. Some formulas have been proposed in the literature to describe $c - I$ and $\eta - I$ relationships, where $\eta = T^s/p_s$ with T^s being the second invariant of τ_s.

Following the work of Boyer et al. [22], Lee et al. [16] assumed

$$c = \frac{c_c}{1 + bI^{1/2}} \tag{55}$$

where c_c is a critical concentration and b is a model parameter. Trulsson et al. [21] proposed

$$\eta = \eta_1 + \frac{\eta_2 - \eta_1}{1 + I_o/I^{1/2}}, \tag{56}$$

where $\eta_1 = \tan\theta_s$ with θ_s being the angle of repose and η_2 and I_o are constants. Based on Eqs. (56) and (55), the following expressions for p_s^r and ν_s^v can be derived [16]:

$$\nu_s^v = \frac{(p_s^r + p_s^e)\eta}{2\rho_s D_s}, \tag{57}$$

which considers the solid phase in its static state as a very viscous fluid and

$$p_s^r = \frac{2b^2 c}{(c_c - c)^2}\left(\rho_f \nu_f + 2a\rho_s d^2 D_s\right) D_s, \tag{58}$$

where b is a constant. In Lee et al. [7], $a = 0.11$ and $b = 1$ were taken.

3.3 Closure models for particle response time

The drag force between the two phases is modeled through the particle response time τ_p. Three representative models for particle response time are introduced in this section.

3.3.1 A model based on the particle sedimentation in still water

The first model is based on particle sedimentation in still water, which can be simplified as a one-dimensional problem, where the steady sedimentation assures that there are no stresses in both the solid and fluid phases in the vertical direction z. In this case, Eqs. (38) and (40) reduce to

$$-\rho_f(1-c)g - (1-c)\frac{\partial p_f}{\partial z} - \frac{c\rho_s(w_f - w_s)}{\tau_p} = 0, \tag{59}$$

and

$$-\rho_s c g - c\frac{\partial p_f}{\partial z} + \frac{c\rho_s(w_f - w_s)}{\tau_p} = 0, \tag{60}$$

where w_f and w_s are the vertical velocities of the fluid and solid phases, respectively.

Because net volume flux through any horizontal plane must be zero, we have

$$(1-c)w_f + cw_s = 0. \tag{61}$$

Combining Eqs. (59) and (61) yields

$$-\frac{\partial p_f}{\partial z} = \frac{-c\rho_s w_s}{(1-c)^2 \tau_p} + \rho_f g. \tag{62}$$

Substituting Eqs. (61) and (62) into Eq. (60) leads to

$$\tau_p = \frac{\rho_s w_s}{(1-c)^2 \left(\rho_s - \rho_f\right) g}, \tag{63}$$

where the solid-phase velocity w_s is also called the hindered settling velocity [23]. The hindered velocity is smaller than the terminal velocity of a single particle, w_0, due to the influence of volumetric concentration (including many-body hydrodynamic interactions). Richardson and Zaki [24] suggested

$$\frac{w_s}{w_0} = (1-c)^n, \tag{64}$$

where the coefficient n is related to the particle Reynolds number $Re_s = w_0 d/\nu_f$

$$n = \begin{cases} 4.65, & Re_s < 0.2 \\ 4.4Re_s^{-0.33}, & 0.2 \le Re_s < 1 \\ 4.4Re_s^{-0.1}, & 1 \le Re_s < 500 \\ 2.4, & 500 \le Re_s \end{cases}. \tag{65}$$

The terminal velocity of a single particle w_0 can be computed by

$$w_0 = \sqrt{\frac{4dg}{3C_d}\frac{\rho_s - \rho_f}{\rho_f}}, \tag{66}$$

where C_d is the drag coefficient for steady flows passing a single particle [25, 26]. For spheres, the following formula of White [27] can be used:

$$C_d = \frac{24}{Re_p} + \frac{6}{1+\sqrt{Re_p}} + 0.4, \tag{67}$$

where $Re_p = |\mathbf{u}_f - \mathbf{u}_s| d/\nu_f$. Combing Eqs. (63)–(67) yields

$$\tau_p = \frac{\rho_s}{\rho_f}\frac{d^2}{\nu_f}\frac{(1-c)^{n-2}}{18 + \left(4.5/\left(1+\sqrt{Re_p}\right) + 0.3\right)Re_p}. \tag{68}$$

It is remarked that Eq. (64) is validated only for $c < 0.4$ [28]. When the concentration c is so high that contact networks form among particles, w_s, becomes zero; when this happens, Eq. (64) is no longer valid any more.

3.3.2 A model based on the pressure drop in steady flows through a homogeneous porous media

Another model for particle response time can be derived by examining the pressure drop in the steady flow through a porous media. For a one-dimensional problem of a horizontal, steady flow through porous media, the terms containing the stresses of the fluid phase disappear, and Eq. (38) reduces to

$$-\frac{\partial p_f}{\partial x} = \frac{c\rho_s u_f}{(1-c)\tau_p}, \tag{69}$$

where the horizontal coordinate x points in the direction of the flow and u is the velocity component in x-direction.

For this problem, Forchheimer [29] suggested

$$-\frac{\partial p_f}{\partial x} = a_F \rho_f (1-c) u_f + b_F \rho_f (1-c)^2 u_f^2, \tag{70}$$

where a_F and b_F are two model parameters. Several formulas for computing a_F and b_F can be found in previous studies. The following two expressions for a_F and b_F suggested by Engelund [25] are recommended for the applications presented at the end of this chapter:

$$a_F = \frac{a_E c^3 \nu_f}{(1-c)^2 d^2}, \quad b_F = \frac{b_E c}{g(1-c)^3 d}, \tag{71}$$

Comparing Eqs. (69) and (70) and using Eq.(71) give

$$\tau_p = \frac{\rho_s d^2}{\rho_f \nu_f} \frac{1}{a_E c^2 + b_E \mathrm{Re}_p}, \tag{72}$$

where a_E and b_E are two model parameters depending on the composition of the solid phase. The parameter a_E is associated with k_p as will be shown later. For $d \approx 2 \times 10^{-4}$ m, $k_p \approx 10^{-10} \sim 10^{-11} \mathrm{m}^2$ [30], which gives $a_E \approx 1.6 \times 10^3 \sim 1.6 \times 10^4$ for $c = 0.5$. The parameter b_E varies from 1.8 to 3.6 or more [28, 31, 32].

For flow in a porous media, the particle response time can also be related to its permeability κ_p. According to Darcy's law for seepage [29], the pressure gradient can also be written as

$$-\frac{\partial p_f}{\partial x} = \frac{\rho_f \nu_f (1-c) u_f}{\kappa_p}, \tag{73}$$

where κ_p is the permeability. Combining Eqs. (69) and (73) gives

$$\tau_p = \frac{c\rho_s \kappa_p}{(1-c)^2 \rho_f \nu_f} \tag{74}$$

When the flow is very slow, Eqs. (70), (71), and (73) suggest that

$$a_E = \frac{d^2}{k_p (1-c)^2}, \tag{75}$$

which means that the particle response time can be related to the permeability.

3.3.3 A hybrid model

Equation (64) is validated only for $c < 0.4$ [28]. To extend Eq. (64) to high concentration regions, Camenen [33] modified Eq. (64) to

$$\frac{w}{w_s} = (1-c)^{n-1}[\max(1-c/c_m, 0)]^{c_m}, \tag{76}$$

where c_m is the maximum concentration at which $w = 0$. In this study, $c_m = c_o$ is adopted because when $c \geq c_o$, contact networks can form in the granular material.

Combining Eqs. (63), (76), and (66)–(67) gives

$$\tau_p = \frac{\rho_s}{\rho_f}\frac{d^2}{\nu_f}\frac{(1-c)^{n-3}[\max(1-c/c_m, 0)]^{c_m}}{18 + \left(4.5/\left(1+\sqrt{Re_p}\right) + 0.3\right)Re_p}. \tag{77}$$

We stress that $c = c_m$ will lead to $\tau_p = 0$ and thus an infinite drag force. Physically, when the volumetric concentration is greater than some critical value, say c_r, Eq. (63) ceases to be valid, and Eq. (72) should be used. To avoid unnaturally large drag force between the two phases, we propose the following model for particle response time:

$$\tau_p = \begin{cases} \dfrac{\rho_s}{\rho_f}\dfrac{d^2}{\nu_f}\dfrac{(1-c)^{n-3}[\max(1-c/c_m, 0)]^{c_m}}{18 + \left(4.5/\left(1+\sqrt{Re_p}\right) + 0.3\right)Re_p}, & \text{for } c < c_r \\ \dfrac{\rho_s d^2}{\rho_f \nu_f}\dfrac{1}{a_E c^2 + b_E Re_p}, & \text{for } c \geq c_r \end{cases} \tag{78}$$

where c_r is the concentration at the intercept point of Eq. (72) and Eq. (77). The transition from Eq. (77) to Eq. (72) is continuous at the intercept point where $c = c_r$. The concentration at the point joining the two models (c_r) is problem-dependent and can be found in principle by solving the following equation:

$$\frac{(1-c_r)^{n-3}[\max(1-c_r/c_m, 0)]^{c_m}}{18 + \left(4.5/\left(1+\sqrt{Re_p}\right) + 0.3\right)Re_p} = \frac{1}{a_E c_r^2 + b_E Re_p}. \tag{79}$$

For given values of a_E and b_E, Eq. (79) implicitly defines c_r as a function of Re_p.

4. Numerical implementation with OpenFOAM

4.1 Introduction to OpenFOAM

This section introduces how to use OpenFOAM® to solve the governing equations with the closure models presented in the previous section. OpenFOAM® is a C++ toolbox developed based on the finite-volume method; it allows CFD code developers to sidestep the discretization of derivative terms on unstructured grids.

4.2 Semidiscretized forms of the governing equations

To avoid numerical noises occurring when $c \to 0$, Rusche [34] suggests that the momentum equations (Eqs. (38) and (40)) should be converted into the following "phase-intensive" form by dividing $\rho_f(1-c)$ and $\rho_s c$:

$$\frac{\partial \mathbf{u}_f}{\partial t} + \nabla\cdot(\mathbf{u}_f\mathbf{u}_f) - (\nabla\cdot\mathbf{u}_f)\mathbf{u}_f = \mathbf{g} - \frac{1}{\rho_f}\nabla p_f + \frac{1}{\rho_f}\nabla\cdot\tau_f - \frac{\tau_f\cdot\nabla c}{\rho_f(1-c)} - \frac{c\rho_s}{\rho_f(1-c)}\frac{(\mathbf{u}_f-\mathbf{u}_s)}{\tau_p} + \frac{\rho_s}{\rho_f(1-c)\tau_p}\frac{\nu_f^t}{\sigma_c}\nabla c \tag{80}$$

and

$$\frac{\partial \mathbf{u}_s}{\partial t} + \nabla\cdot(\mathbf{u}_s\mathbf{u}_s) - (\nabla\cdot\mathbf{u}_s)\mathbf{u}_s = \mathbf{g} - \frac{1}{\rho_s}\nabla p_f - \frac{1}{\rho_s c}\nabla c p_s + \frac{1}{\rho_s}\nabla\cdot\tau_s + \frac{\tau_s\cdot\nabla c}{\rho_s c} + \frac{(\mathbf{u}_f-\mathbf{u}_s)}{\tau_p} - \frac{1}{c\tau_p}\frac{\nu_f^t}{\sigma_c}\nabla c \tag{81}$$

The solutions of Eqs. (80) and (81) are expressed in the following semidiscretized forms:

$$\mathbf{u}_f = \frac{\mathbf{A}_H^f}{\mathbf{A}_D^f} + \frac{\mathbf{g}}{A_D^f} - \frac{\nabla p_f}{\rho_f \mathbf{A}_D^f} + \frac{\rho_s c \mathbf{u}^s}{\rho_f \mathbf{A}_D^f(1-c)\tau_p} + \frac{\rho_s}{\rho_f \mathbf{A}_D^f(1-c)\tau_p}\frac{\nu_f^t}{\sigma_c}\nabla c \tag{82}$$

$$\mathbf{u}_s = \frac{\mathbf{A}_H^s}{\mathbf{A}_D^s} + \frac{\mathbf{g}}{\mathbf{A}_D^s} - \frac{\nabla p_f}{\rho_s A_D^s} - \frac{\nabla p_s}{\rho_s \mathbf{A}_D^s} - \frac{p_s\nabla c}{\rho_s \mathbf{A}_D^s c} + \frac{\rho_s \mathbf{u}_f}{\mathbf{A}_D^s \tau_p} - \frac{1}{\mathbf{A}_D^s c\tau_p}\frac{\nu_f^t}{\sigma_c}\nabla c \tag{83}$$

where $\mathbf{A}^\beta$ ($\beta = s$ or f) denotes the systems of linear algebraic equations arising from the discretization of either Eqs. (82) or (83). The matrix $\mathbf{A}^\beta$ is decomposed into a diagonal matrix, $\mathbf{A}_D^\beta$, and an off-diagonal matrix, $\mathbf{A}_O^w$. Also, $\mathbf{A}_H^w = \mathbf{b}^w - \mathbf{A}_O^\beta\mathbf{u}^\beta$ with $\mathbf{b}^\beta$ relating to the second to final terms on the right-hand side of either Eqs. (82) or (83). OpenFOAM® built-in functions are used to compute $\mathbf{A}_D^\beta$ and $\mathbf{A}_H^\beta$, which depend on the discretization schemes. For example, Lee et al. [16] and Lee and Huang [35] used a second-order time-implicit scheme and a limited linear interpolation scheme for all variables except for velocity. To interpolate velocities, the total-variation-diminishing (TVD) limited linear interpolation scheme is adopted for velocity.

4.3 A prediction-correction method

If Eq. (83) is directly used to calculate $\mathbf{u}_s$ and Eq. (39) to calculate c, then c may increase rapidly toward c_c, leading to an infinite p_s for large c. This can be avoided by using a prediction-correction method to compute $\mathbf{u}_f$ and $\mathbf{u}_s$. This is achieved by splitting Eq. (83) into a predictor $\mathbf{u}_s^*$ and a corrector. The predictor is

$$\mathbf{u}_s^* = \frac{\mathbf{A}_H^s}{\mathbf{A}_D^s} + \frac{\mathbf{g}}{\mathbf{A}_D^s} - \frac{\nabla p_f}{\rho_s A_D^s} + \frac{\rho_s \mathbf{u}_f}{\mathbf{A}_D^s \tau_p} \tag{84}$$

which is corrected by the following corrector

$$\mathbf{u}_s = \mathbf{u}_s^* - \left(\frac{p_s\nabla c}{\rho_s \mathbf{A}_D^s c} - \frac{1}{\mathbf{A}_D^s c\tau_p}\frac{\nu_f^t}{\sigma_c}\nabla c\right) \tag{85}$$

This predictor-corrector scheme can improve the numerical stability by introducing a numerical diffusion term. To see this, we combine Eqs. (39) and (85) to obtain the following equation describing the evolution of c:

$$\frac{\partial c}{\partial t} + \nabla \cdot \left(c\mathbf{u}_s'\right) = \nabla \cdot \left(\frac{p_s}{\rho_s \mathbf{A}_D^s} + \frac{1}{\mathbf{A}_D^s \tau_p} \frac{\nu_f^t}{\sigma_c}\right) \nabla c \tag{86}$$

The right-hand side of Eq. (86) now has a diffusive term introduced by the numerical scheme. High sediment concentration and large p_s increase the numerical diffusion (the right-hand side of Eq. (86)) and thus can avoid a rapid increase of c and the numerical instability due to high sediment concentration.

For the velocity-pressure coupling, Eq. (82) is similarly solved using a predictor $\mathbf{u}_f^*$ and a corrector. The predictor is

$$\mathbf{u}_f^* = \frac{\mathbf{A}_H^f}{\mathbf{A}_D^f} + \frac{\mathbf{g}}{A_D^f} + \frac{\rho_s c \mathbf{u}_s}{\rho_f \mathbf{A}_D^f (1-c)\tau_p} + \frac{\rho_s}{\rho_f \mathbf{A}_D^f (1-c)\tau_p} \frac{\nu_f^t}{\sigma_c} \nabla c \tag{87}$$

which is corrected by the following corrector

$$\mathbf{u}_f = \mathbf{u}_f^* - \frac{\nabla p_f}{\rho_f \mathbf{A}_D^f} \tag{88}$$

Substituting Eq. (88) into Eq. (37) gives a pressure equation. However, when using this pressure equation to simulate air-water flows, numerical experiments have shown that the lighter material is poorly conserved [36]. The poor conservation of lighter material can be avoided by combining Eqs. (37) and (39) into the following Eq. (37):

$$\nabla \cdot \left[(1-c)\mathbf{u}_f + c\mathbf{u}_s\right] = 0 \tag{89}$$

and using Eq. (89) to correct p_f. The method proposed [37] can help avoid the numerical instability. To show this, we follow Carver [37] and define

$$\hat{\mathbf{u}}_s = \frac{\mathbf{A}_H^s}{\mathbf{A}_D^s} + \frac{\mathbf{g}}{\mathbf{A}_D^s} - \frac{\nabla p_s}{\rho_s \mathbf{A}_D^s} - \frac{p_s \nabla c}{\rho_s \mathbf{A}_D^s c} + \frac{\rho_s \mathbf{u}_f}{\mathbf{A}_D^s \tau_p} - \frac{1}{\mathbf{A}_D^s c \tau_p} \frac{\nu_f^t}{\sigma_c} \nabla c \tag{90}$$

and combine Eqs. (83) and (88)–(90) to obtain the following equation

$$\nabla \cdot \left[(1-c)\hat{\mathbf{u}}_f + c\hat{\mathbf{u}}_s\right] = \nabla \cdot \left[\frac{1-c}{\rho_f \mathbf{A}_D^f} + \frac{c}{\rho_s \mathbf{A}_D^s}\right] \nabla p_f \tag{91}$$

The numerical diffusion term on the right-hand side of Eq. (91) can help improve the numerical stability.

The prediction-correction method presented here deals with velocity-pressure coupling and avoids the numerical instability caused by high concentration. The turbulence closure $k - \epsilon$ model is also solved in "phase-intensive" forms. For other details relating to the numerical treatments, the reader is referred to "twoPhaseEulerFoam," a two-phase solver provided by OpenFOAM®.

4.3.1 Outline of the solution procedure

When $c \to 0$, Eq. (83) becomes singular. To avoid this, $1/c$ is replaced by $1/(c + \delta c)$ in numerical computations, where δc is a very small number, say 10^{-6}. When $c \leq \delta c$, only a very small amount of solid particles are moving with the fluid; replacing $1/c$ by $1/(c + \delta c)$ may introduce error in computing $\mathbf{u}^s$; to avoid this error, we can set $\mathbf{u}_s = \mathbf{u}_f$, which means the solid particles completely follow the water particles; this does not affect the computations of other variables because the momentum of the solid phase $c\mathbf{u}_s$ is very small when $c \leq 10^{-6}$. Because the maximum value of c is always smaller than 1, there is no singularity issue with Eq. (82).

An iteration procedure is needed to solve the governing equations at each time step for the values of c,$\mathbf{u}_f$, $\hat{\mathbf{u}}_s$, and p_f obtained at the previous time step, and it is outlined below:

1. Solve Eqs. (80) and (81).
2. Compute $\mathbf{u}_s^*$ from Eq. (84).
3. Solve Eq. (86) for c.
4. Compute $\mathbf{u}_s$ from Eq. (85).
5. Compute $\hat{\mathbf{u}}_s$ from Eq. (90).
6. Compute $\mathbf{u}_f^*$ from Eq. (87).
7. Solve Eq. (91) for p_f.
8. Repeat Eqs. (5)–(7) for n times (say $n = 1$).
9. Compute $\mathbf{u}_f$ from Eq. (88).
10. Set $\mathbf{u}_s = \mathbf{u}_f$ for very dilute region, specifically $c \leq 10^{-6}$.
11. Repeat Eqs. (1)–(10) with the updated c, $\mathbf{u}_f$, $\hat{\mathbf{u}}_s$, and p_f until the residuals of Eqs. (80), (86), and (91) are smaller than the tolerance (say 10^{-5}).
12. Solve Eqs. (45) and (46) for k and ϵ, and compute the related coefficients.

Figure 1 is a flowchart showing these 12 solution steps.

In the absence of the solid phase, the numerical scheme outlined here reduces to the "PIMPLE" scheme, which is a combination of the "pressure implicit with splitting of operator" (PISO) scheme and the "semi-implicit method for pressure-linked equations" (SIMPLE) scheme. Iterations need to be done separately to solve Eq. (80) for $\mathbf{u}_f$, Eq. (81) for $\mathbf{u}_s$, Eq. (86) for c, Eq. (91) for p_f, Eq. (45) for k, and (46) for ϵ; the convergence criteria are set at the residuals not exceeding 10^{-8}. Because Eqs. (80), (81), (86), and (87) are coupled, additional residual checks need to be performed at step 11; however, the residual for Eq. (81) is not checked because $\mathbf{u}_s = \mathbf{u}_f$ is enforced in step 10.

To ensure the stability of the overall numerical scheme, the Courant-Friedrichs-Lewy (CFL) condition must be satisfied for each cell. The local Courant number for each cell, which is related to the ratio between the distance of a particle moving within Δt and the size of the cell where such particle is located, is defined as

$$\mathrm{CFL} = \sum \frac{\mathrm{abs}\left(\mathbf{u}^j \cdot \mathbf{S}^j\right)}{2V} \Delta t, \tag{92}$$

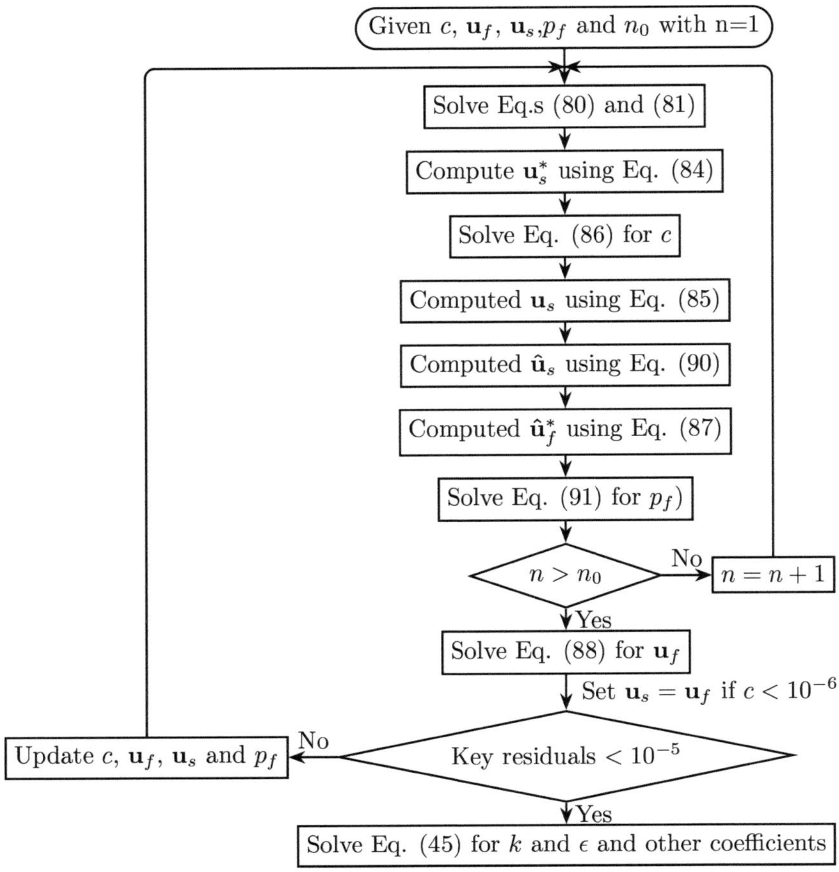

Figure 1.
A flow chart showing the solution procedure using OpenFOAM®.

where in $\mathbf{u}^j = (1-c)\mathbf{u}_f^j + c\mathbf{u}_s^j$, the subscript "$j$" represents the jth face of the cell, $\mathbf{S}^j$ is a unit normal vector, V is the volume of the cell, and Δt is the time step. The Courant number must be less than 1 to avoid numerical instability. Generally, max (CFL) <0.1 is suggested. The values of CFL for high concentration regions should be much smaller than those for low concentration regions so that rapid changes of c can be avoided. Therefore, it is recommended that $\max\left(\text{CFL}|_{c>c_o}\right) < 0.005$. The time step is recommended to be in the range of 10^{-5} and 10^{-4} s.

5. Applications

This section briefly describes two examples that have been studied using the two-phase flow models described. The problem descriptions and numerical setups for these two problems are included here; for other relevant information, the reader is referred to Lee and Huang [35] and Lee et al. [38].

5.1 Scour downstream of a sluice gate

A sluice gate is a hydraulic structure used to control the flow in a water channel. Sluice gate structures usually have a rigid floor followed by an erodible bed. The scour downstream of a sluice gate is caused by the horizontal submerged water jet issuing from the sluice gate. It is of practical importance to understand the maximum scour depth for the safety of a sluice gate structure. Many experimental studies have been done to investigate the maximum scour depth and the evolution of scour profile (e.g., Chatterjee et al. [39]). For numerical simulations, this problem includes water (fluid phase) and sediment (solid phase) and is best modeled by a liquid-solid two-phase flow approach. In the following, the numerical setup and main conclusions used in Lee et al. [38] are briefly described. The experimental setup of Chatterjee et al. [39] is shown in **Figure 2**. To numerically simulate the experiment of [8], we use the same sand and dimensions to set up the numerical simulations: quartz sand with $\rho_s = 2650$ kg/m^3 and $d = 0.76$ mm is placed in the sediment reservoir, with its top surface being on the same level as the top surface of the apron; the sluice gate opening is 2 cm; the length of apron is 0.66 m; the sediment reservoir length is 2.1 m; the overflow weir on the right end has a height of 0.239 m; the upstream inflow discharge rate at the sluice opening is 0.204 m^2/s, which translates into an average horizontal flow velocity $V = 1.02$ m/s under the sluice gate. As an example, the computed development of scour depth d_s is shown in **Figure 3** together with the measurement of Chatterjee et al. [39].

The problem involves also an air-water surface, which can be tracked using a modified volume-of-fluid method introduced in [38]. A nonuniform mesh is used in the two-phase flow simulation because of the air-water interface, the interfacial momentum transfer at the bed, and the large velocity variation due to the water jet. The finest mesh with a vertical mesh resolution of $2d$ is used in the vicinity of the sediment-fluid interface; this fine mesh covers the dynamic sediment-fluid

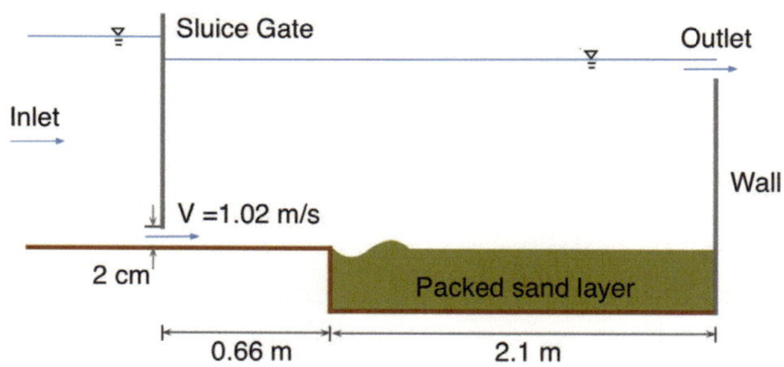

Figure 2.
A sketch of the experimental setup for scour induced by a submerged water jet.

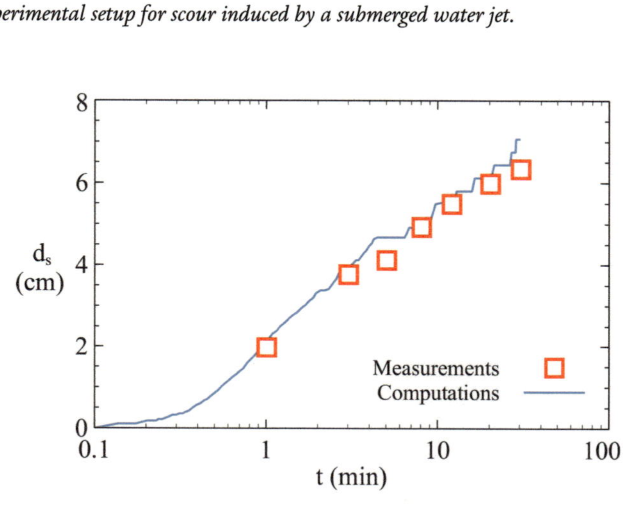

Figure 3.
Comparison of the computed scour depth with measurements of Chatterjee et al. [39].

interface during the entire simulation. In regions away from the sediment-fluid interface or regions where the scouring is predicted to be negligible (e.g., further downstream the scour hole), the mesh sizes with a vertical resolution ranging from 3 to 5 mm are used. The aspect ratio of the mesh outside the wall jet region is less than 3.0. Since in the wall jet, horizontal velocity is significantly larger than the vertical velocity, the aspect ratio of the local mesh in the wall jet region is less than 5.0.

The scour process is sensitive to the model for particle response time used in the simulation. Because Eq. (72) can provide a better prediction of sediment transport rate for small values of Shields parameter, it is recommended for this problem. The two-phase flow model can reproduce well the measured scour depth and the location of sand dune downstream of the scour hole.

5.2 Collapse of a deeply submerged granular column

Another application of the fluid-solid two-phase flow simulation is the simulation of the collapse of a deeply submerged granular column. The problem is best described as a granular flow problem, which involves sediment (a solid phase) and water (fluid phase). Many experimental studies have been reported in the literature on this topic. This section describes a numerical simulation using the fluid-solid two-phase flow model described in this chapter.

Figure 4 shows the experimental setup of Rondon et al. [40]. A 1:1 scale two-phase flow simulation was performed by Lee and Huang [35] using the fluid-solid two-phase flow model presented in this chapter. The diameter and the density of the sand grain are 0.225 mm and 2500 kg/m^3, respectively. The density and the dynamic viscosity of the liquid are 1010 kg/m^3 and 12 mPa s, respectively. Note that the viscosity of the liquid in the experiment is ten times larger than that for water at room temperature. For this problem, using a mesh of 1.0 $\times$ 1.0 mm and the particle response model given by Eq. (78), the fluid-solid two-phase flow model presented in this chapter can reproduce well the collapse process reported in Rondon et al. [40]. **Figure 5** shows the simulated collapsing processes compared with the measurement for two initial packing conditions: initially loosely packed condition and initially densely packed condition.

The two-phase model and closure models presented in this chapter are able to deal with both initially loose packing and initially dense packing conditions and reveal the roles played by the contractancy inside the granular column with a loose packing and dilatancy inside a granular column with a dense packing. One of the conclusions of Lee and Huang [35] is that the collapse process of a densely packed granular column is more sensitive to the model used for particle response time than that of a loosely packed granular column. The particle response model given by Eq. (78) performs better than other models; this is possibly because the liquid used in Rondon et al. [40] is much viscous than water.

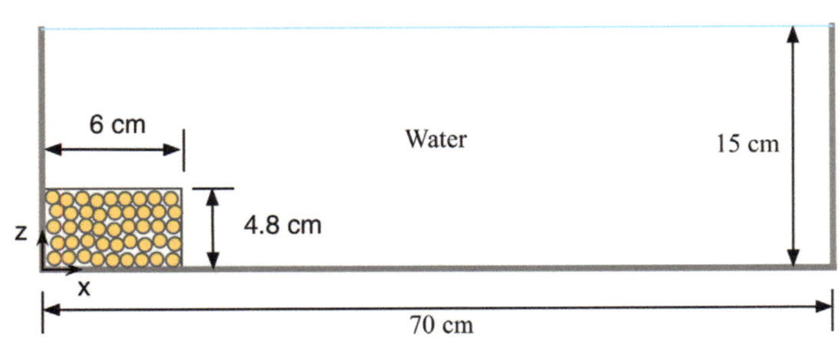

Figure 4.
A sketch of the experimental setup for the collapse of a deeply submerged granular column.

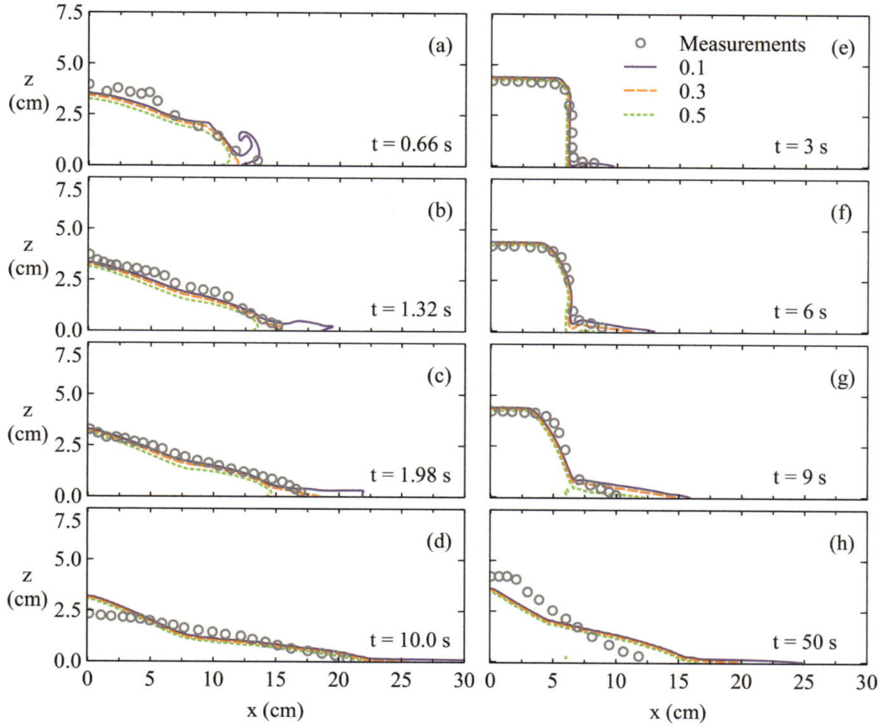

Figure 5.
The simulated collapsing processes for the initially loose condition (a)–(d) and the initially dense condition (e)–(h). The lines represent contours of the computed concentrations, and the symbols were experimental data of Rondon et al. [40]. The figure is adapted from Lee and Huang [35].

6. Summary

This chapter presented a brief introduction to the equations and closure models suitable for fluid-solid two-phase flow problems such as sediment transport, submarine landslides, and scour at hydraulic structures. Two averaging operations were performed to derive the governing equations so that the turbulent dispersion, important for geophysical flow problems, can be considered. A new model for the rheological characteristics of sediment phase was used when computing the stresses of the solid phase. The $k - \epsilon$ model was used to determine the Reynolds stresses. A hybrid model to compute the particle response time was introduced, and the numerical implementation in the framework of OpenFOAM® was discussed. A numerical scheme was introduced to avoid numerical instability when the concentration is high. Two applications were describe to show the capacity of the two-phase flow models presented in this chapter.

Acknowledgements

This material presented here is partially based upon work supported by the National Science Foundation under Grant No. 1706938 and the Ministry of Science and Technology, Taiwan [MOST 107-2221-E-032-018-MY3]. Any opinions, findings, and conclusions or recommendations expressed in this material are those of the author(s) and do not necessarily reflect the views of the National Science Foundation.

Author details

Zhenhua Huang[1]* and Cheng-Hsien Lee[2]

1 Department of Ocean and Resources Engineering, School of Ocean and Earth Science and Technology, University of Hawaii at Manoa, Honolulu, HI, USA

2 Department of Water Resources and Environmental Engineering, Tamkang University, New Taipei City, Taiwan

*Address all correspondence to: zhenhua@hawaii.edu

References

[1] Balachandar S, Eaton JK. Turbulent dispersed multiphase flow. Annual Review of Fluid Mechanics. 2010;**42**: 111-133

[2] Picano P, Breugem WP, Brandt L. Turbulent channel flow of dense suspensions of neutrally buoyant spheres. Journal of Fluid Mechanics. 2015;**764**:463-487

[3] Vreman AW. Turbulence attenuation in particle-laden flow in smooth and rough channels. Journal of Fluid Mechanics. 2015;**73**: 103-136

[4] Hsu TJ, Liu PLF. Toward modeling turbulent suspension of sand in the nearshore. Journal of Geophysical Research. 2004;**109**:C06018

[5] Jha SK, Bombardelli FA. Toward two-phase flow modeling of nondilute sediment transport in open channels. Journal of Geophysical Research. 2010; **115**:F03015

[6] Jha SK, Bombardelli FA. Theoretical/numerical model for the transport of non-uniform suspended sediment in open channels. Advances in Water Resources. 2011;**34**:577-591

[7] Lee CH, Huang ZH, Chiew YM. A multi-scale turbulent dispersion model for dilute flows with suspended sediment. Advances in Water Resources. 2015;**79**:18-34

[8] Drew DA. Mathematical modeling of two-phase flow. Annual Review of Fluid Mechanics. 1983;**15**(1): 261-291

[9] Hsu TJ, Jenkins JT, Liu PLF. On two-phase sediment transport: Dilute flow. Journal of Geophysical Research. 2003; **108**:C33057

[10] Chandrasekharaiah DS, Debnath L. Continuum Mechanics. California: Academic Press; 1994

[11] Hwang GJ, Shen HH. Modeling the phase interaction in the momentum equations of a fluid solid mixture. International Journal of Multiphase Flow. 1991;**17**(1):45-57

[12] Drew DA. Turbulent sediment transport over a flat bottom using momentum balance. Journal of Applied Mechanics. 1975;**42**(1):38-44

[13] Revil-Baudard T, Chauchat J. A two phase model for sheet flow regime based on dense granular flow rheology. Journal of Geophysical Research. 2013; **118**:1-16

[14] Chiodi F, Claudin P, Andreotti B. A two-phase flow model of sediment transport: Transition from bedload to suspended load. Journal of Fluid Mechanics. 2014;**755**:561-581

[15] Launder BE, Sharma BI. Application of the energy-dissipation model of turbulence to the calculation of flow near a spinning disc. Letters in Heat and Mass Transfer. 1974;**1**(2):131-137

[16] Lee CH, Low YM, Chiew YM. Multi-dimensional rheology-based two-phase model for sediment transport and applications to sheet flow and pipeline scour. Physics of Fluids. 2016;**28**:053305

[17] Song C, Wang P, Makse HA. A phase diagram for jammed matter. Nature. 2008;**453**:629-632

[18] Crowe CT. On models for turbulence modulation in fluid–particle flows. International Journal of Multiphase Flow. 2000;**26**(5):719-727

[19] Hsu TJ, Jenkins JT, Liu PLF. On two-phase sediment transport: Sheet

flow of massive particles. Proceedings of the Royal Society A: Mathematical, Physical and Engineering Science. 2004; **460**(2048):2223-2250

[20] Hinze J. Turbulence. New York: McGraw Hill; 1959

[21] Trulsson M, Andreotti B, Claudin P. Transition from the viscous to inertial regime in dense suspensions. Physical Review Letters. 2012;**109**(11):118305

[22] Boyer F, Guazzelli É, Pouliquen O. Unifying suspension and granular rheology. Physical Review Letters. 2011; **107**(18):188301

[23] Pitman EB, Le L. A two-fluid model for avalanche and debris flows. Philosophical Transactions. Series A, Mathematical, Physical, and Engineering Sciences. 2005;**363**(1832): 1573-1601

[24] Richardson JF, Zaki WN. Sedimentation and fluidisation: Part I. Chemical Engineering Research and Design. 1954;**32**:S82-S100

[25] Engelund F. On the Laminar and Turbulent Flows of Ground Water Through Homogeneous Sand. Copenhagen: Danish Academy of Technical Sciences; 1953

[26] Chien N, Wan Z. Mechanics of Sediment Transport. Reston: American Society of Civil Engineers; 1999

[27] White FM. Viscous Fluid Flow. Singapore: McGraw-Hill; 2000

[28] Yin X, Koch DL. Hindered settling velocity and microstructure in suspensions of solid spheres with moderate Reynolds numbers. Physics of Fluids. 2007;**19**:093302

[29] Bear J. Dynamics of Fluids in Porous Media. New York: American Elsevier; 1972

[30] Das BM. Principles of Geotechnical Engineering. Stamford: Cengage Learning; 2013

[31] Burcharth HF, Andersen OH. On the one-dimensional steady and unsteady porous flow equations. Coastal Engineering. 1995;**24**:233-257

[32] Higuera P, Lara JL, Losada IJ. Three-dimensional interaction of waves and porous coastal structures using OpenFOAM (R). Part I: Formulation and validation. Coastal Engineering. 2014;**83**:243-258

[33] Camenen B. Settling velocity of sediments at high concentrations. Coastal Engineering. 2005;**51**(1):91-100

[34] Rusche H. Computational fluid dynamics of dispersed two-phase flows at high phase fractions [PhD thesis]. London: University of London; 2003

[35] Lee CH, Huang ZH. A two-phase flow model for submarine granular flows: With an application to collapse of deeply-submerged granular columns. Advances in Water Resources. 2018;**115**: 286-300

[36] Hancox WT, Banerjee S. Numerical standards for flow-boiling analysis. Nuclear Science and Engineering. 1977; **64**(1):106-123

[37] Carver MB. Numerical computation of phase separation in two fluid flow. Journal of Fluids Engineering. 1984; **106**(2):147-153

[38] Lee CH, Xu CH, Huang ZH. A three-phase flow simulation of local scour caused by a submerged wall-jet with a water-air interface. Advances in Water Resources. 2018. DOI: 10.1016/j.advwaters.2017.07.017. In Press

[39] Chatterjee SS, Ghosh SN, Chatterjee M. Local scour due to submerged horizontal jet. Journal of Hydraulic Engineering. 1994;**120**(8):973-992

[40] Rondon L, Pouliquen O, Aussillous P. Granular collapse in a fluid: Role of the initial volume fraction. Physics of Fluids. 2004;**23**:73301

Chapter 7

CFD Simulation of Flow Phenomena in Selected Centrifugal Pumps, Industrial Fans and Positive Displacement Pumps

Wieslaw Fiebig, Paulina Szwemin and Maciej Zawislak

Abstract

The chapter presents simulation models for the analysis of centrifugal pumps, fans and positive displacement pumps. In centrifugal pumps based on the "sliding mesh" method, a CFD model was created to calculate the flow characteristics, and the pump operating parameters were determined at which an unfavourable phenomenon of cavitation occurs. In the case of a radial fan, the CFD model was used to determine the influence of inlet channel geometry on the efficiency of an industrial installation. The main purpose of the CFD simulation was to obtain the pressure distributions and determine the areas in which cavitation may occur. To investigate the flow phenomena that occur in external gear pumps and double-acting vane pumps, the "immersed solid" method was used. The results of 2D and 3D simulation studies for various operating parameters of pumps have been presented.

Keywords: CFD simulation, pumps, cavitation, industrial fans, flow analysis

1. Introduction

Industrial machines and devices with rotating operating parts are difficult to model due to their complex geometry, the transition of elements of the discrete model between the rotating and non-rotating parts, the importance of the quality of elements of the discrete model, and the fact that in most cases, it is necessary to take into account the time step (elements rotate in relation to the casing). It is also troublesome that very often the calculations are stabilised only after a few rotations of the operating element. However, the use of computational fluid dynamics methods to model this group of machines and equipment is justified, as it enables:

- Determining the internal and external characteristics of machines and devices in virtual space

- Imaging and observing the flow phenomena in the machine itself (especially when for various reasons it is impossible to measure physical quantities of the flowing medium)

- Designing equipment for which there are no design guidelines (e.g. differentials, mixers)
- Improving the efficiency of machinery and equipment

In this chapter, selected examples of numerical calculations will be described, showing the possibility of using CFD methods to solve machine and equipment problems with a rotating operating element, often found in industrial practice.

2. Vane pump: flow analysis

The innovative vane pump described in study [1] was subjected to the analysis of flow phenomena. In this solution, the pump is integrated into the BLDC permanent magnet electric motor. Due to its design, which differs from the standard solutions, it was necessary to check whether cavitation could occur in the suction channel of the pump. The main objective of the CFD simulation was to determine the areas where cavitation is likely to occur [2] and its intensity depending on the rotational velocity. The subject of the study is a positive displacement pump with integrated electric drive, consisting of an impeller embedded in a casing. Unlike conventional gear and vane pumps [3–10], the pump impeller and motor stator are immovable components, while the pump casing rotates with the rotor of the electric motor. **Figure 1** shows the 3D model of the analysed pump.

An important problem is to examine the flow in the suction channel of the pump, as it is exposed to the adverse effects of cavitation, which can develop as a result of a too high value of negative pressure occurring in the suction area.

On the basis of the three-dimensional model of the pump, a geometric model of the volume of operating fluid filling its interior was prepared (**Figure 1b**). As expected, the result is a very complex structure in terms of geometry. Due to the particular interest in the phenomena occurring in the suction channel of the pump, the calculations used a fragment of the geometric model of the operating fluid volume filling the interior of the pump, which is the volume of oil filling the pump from the inlet to the suction kidneys supplying the fluid to the inter-vane spaces (**Figure 2a**). The separated volume is contained in the immovable elements of the structure, which further simplifies the formulation of the flow problem and the choice of calculation parameters.

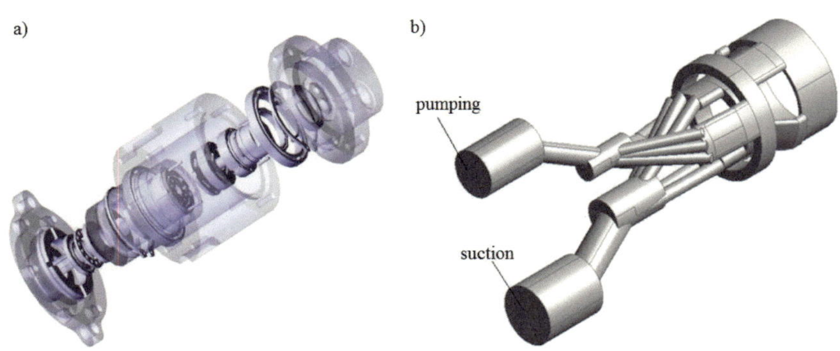

Figure 1.
(a) 3D model of the vane pump with integrated mechatronic electric drive and (b) 3D model of the operating fluid volume filling the pump.

DOI: http://dx.doi.org/10.5772/intechopen.82266

Based on the three-dimensional pump model, the simplified geometric model of the operating fluid volume filling the suction channel was discretized using a tetrahedral grid. The result is a geometric model divided into 144,390 tetrahedral elements with 29,711 nodes, as shown in **Figure 2b**.

The next step in formulating the flow problem is to select the type and define the boundary conditions for relevant fragments of the geometry. In the analysed case, the conditions concerning the fluid inflow and outflow were set as shown in **Figure 2a**.

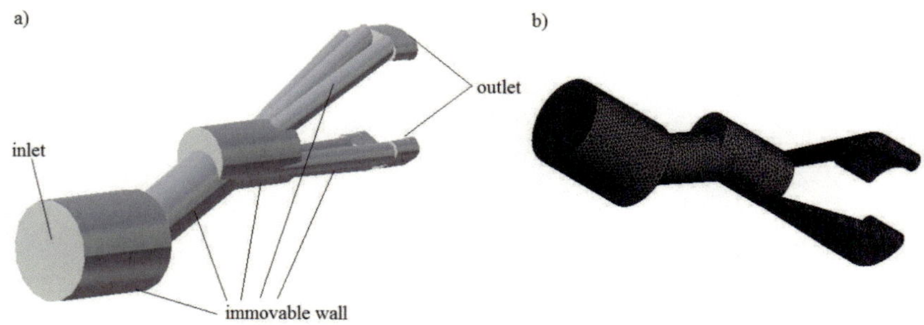

Figure 2.
(a) Suction channel geometry and (b) discrete model.

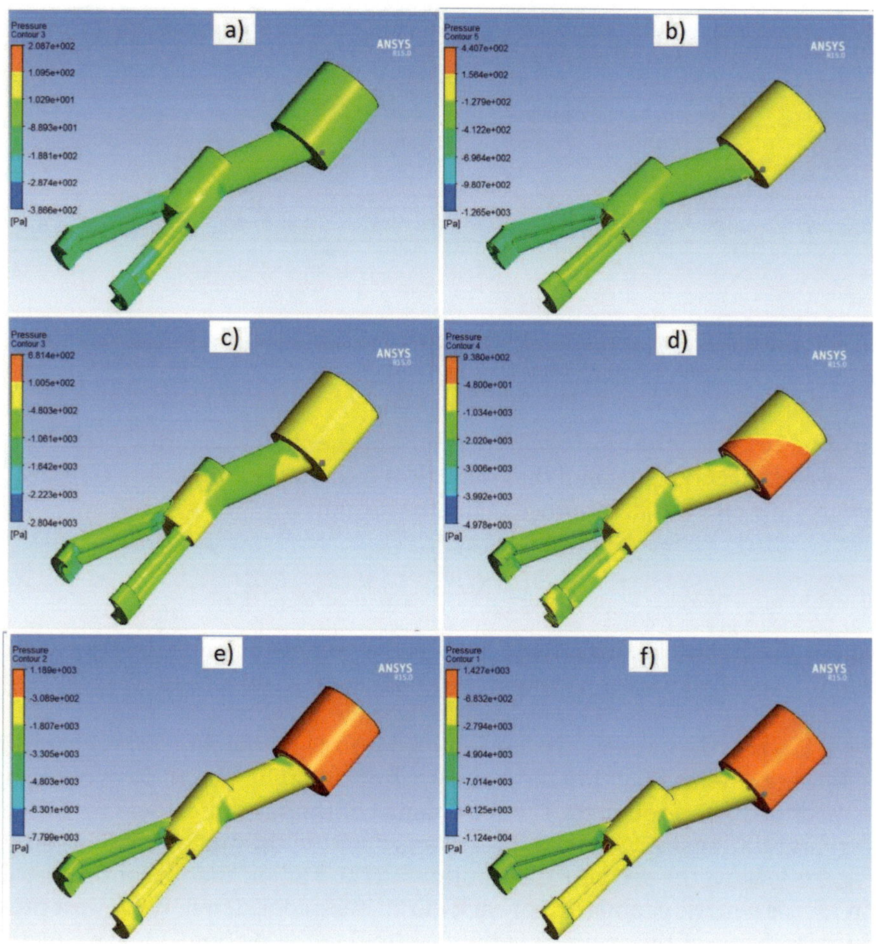

Figure 3.
Generic geometry—suction channel pressure distribution for different rotational velocities: (a) 500 rpm, (b) 1000 rpm, (c) 1500 rpm, (d) 2000 rpm, (e) 2500 rpm and (f) 3000 rpm.

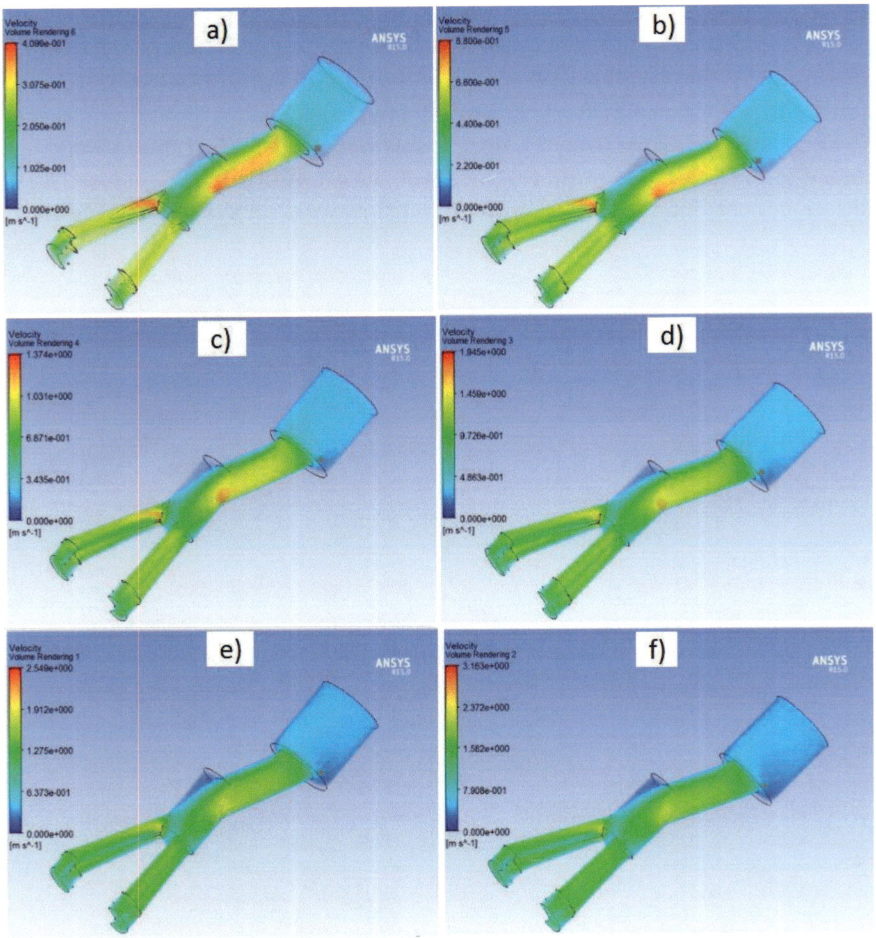

Figure 4.
Generic geometry—suction channel velocity distribution for different velocities: (a) 500 rpm, (b) 1000 rpm, (c) 1500 rpm, (d) 2000 rpm, (e) 2500 rpm and (f) 3000 rpm.

In order to obtain the most accurate results of the simulation, the "pressure inlet" condition at the inlet and the "mass flow rate" at the outlet were assumed. The mass flow rate was determined using the formula:

$$q = 2zb\left[\frac{\pi}{z}\left(R_2^2 - R_1^2\right) - w\left(R_2 - R_1\right)\right] \tag{1}$$

where q is the specific mass flow rate; z is the number of vanes; b is the width of a vane; w is the thickness of a vane; *R1* is the small race radius; and R2 is the large race radius.

On this basis, the numerical values entered into the simulation for each impeller velocity were obtained. Within the framework of the study, the analysis of the operating medium flow through the suction channel of the vane pump was performed for various rotational velocity values—changed within the range of 500–3000 rpm.

Figure 3 shows the pressure distributions in the suction channel of the tested pump for the generic geometry. For each of the cases considered, the lowest pressure occurs in one of the channels supplying fluid to the suction kidneys directly at the inlet to the channel. It was found that the negative pressures for the whole range of rotational velocities are higher than the pressure of oil evaporation, which prevents the occurrence of cavitation phenomena.

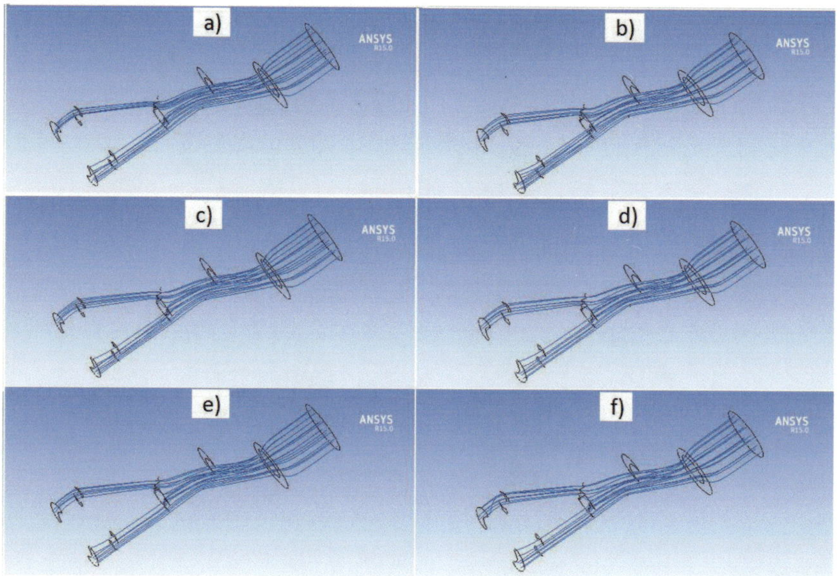

Figure 5.
Generic geometry—streamlines in the investigated suction channel area for different velocities: (a) 500 rpm, (b) 1000 rpm, (c) 1500 rpm, (d) 2000 rpm, (e) 2500 rpm and (f) 3000 rpm.

The results of the calculations, apart from pressure distributions, were presented in the form of velocity distributions in the considered area, which are presented in **Figure 4**. From the obtained velocity distributions, it appears that the rotational velocity of the pump significantly influences the velocity of fluid flow in one of the supply channels for both the generic and the modified geometry. It is worth noting that the area where the highest velocities were identified corresponds to the area of the lowest pressures observed in the suction channel. The velocity of the fluid decreases with the lowering of the rotational velocity, but in the case of simplified geometry, it is slightly lower.

Figure 5 shows the fluid flow in the form of streamlines, for which the inflow plane to the domain is assigned as the beginning. The results obtained confirm the previous assumptions that the fluid flows evenly and without major turbulences through both inlet channels. Uneven velocity distribution and different pressure values due to asymmetrical layout of channels did not affect the fluid flow. The results obtained on the basis of numerical calculations are the basis for evaluation of the structure of channels supplying fluid to the inter-vane volumes.

3. Radial fan: characteristics and performance improvement

Another object under consideration with rotating operating elements was a radial fan. The aim of the numerical simulation was to improve its efficiency. The flow of real gas through a fan with a finite amount of blades is carried out by the cost of loss of energy, called hydraulic losses. Those losses are a consequence of the friction of air molecules occurring on the blade walls and fan housing, vortexes developed in the gas stream, etc. The influence of hydraulic losses on the working characteristic of the radial fan is described by a hydraulic efficiency coefficient, which is defined as the ratio of the useful power to the power delivered by the impeller. This coefficient also defines the real delivery height to the theoretical delivery height—obtained for the finite amount of impeller blades. The impeller geometry considered in possible options, i.e. with eight (factory option) and nine (suggested option) vanes, are shown in **Figures 6** and 7.

Figure 6.
Impeller shape: eight vanes.

Figure 7.
Impeller shape: nine vanes.

Figure 8.
Discrete model with division into tetrahedral elements on the impeller and vanes.

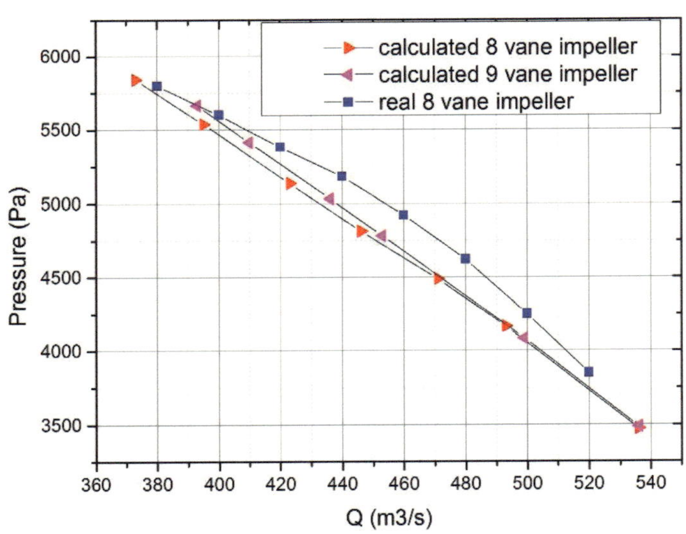

Figure 9.
Comparison of the calculation results and the results of the technical documentation for the impeller with the eight and nine vanes.

Figure 10.
Total velocity [m/s] distribution for calculated impeller operating points with eight and nine vanes.

For the calculations, the model of impeller according to the enclosed documentation was used as the output model. Calculations have been made for both impeller variants. For both of the cases, the discrete model was based on tetrahedral elements (as exemplary shown in **Figure 8**). Elements near walls were compacted. The flow was modelled as turbulent, using the RANS method and the two-equation turbulence model k-ε.

In the first stage of the study, the analysis of the impeller with eight (**Figure 6**) and nine (**Figure 7**) vanes was carried out. For the eight vanes, the results of the simulation were also compared with the available results in the technical documentation and found to be similar (**Figure 9**). Furthermore, the overall performance of the two types of impellers found with aid of CFD calculation maintains in similar level.

In order to verify the correctness of the calculation of the main dimensions of the impeller, a theoretical design process was carried out. On the basis of known designs, the influence of impeller parameters on its performance, compression and efficiency was simulated. It was necessary to maintain the existing parameters of the impeller, improving only its efficiency. The modifications were limited by the external dimensions of the impeller in order to be able to work with the existing collecting volute.

After a number of variant combinations, the outlet angle of the vane was changed to 23° and the vane profile modified to improve efficiency. The results show that by changing the outlet angle, the average efficiency for the eight-vane impeller was increased by 2.3% and for the nine-vane impeller by 2.9% in relation to the basic eight-vane impeller.

Figure 10 shows a comparison of the flow images for the impellers with eight and nine vanes with a 23° outlet angle.

The best results were obtained for the nine-vane impeller and the changed outlet angle. An average efficiency increase of 2.9% was achieved in relation to the impeller from the technical documentation. The flow images are correct. There are no particularly dangerous phenomena, such as interruption of flow or turbulence.

4. Centrifugal pump with collecting channel: undetermined flow with cavitation

Another object of the study was a single-stage centrifugal pump with a spiral volute cooperating with two similar types of impellers, commonly used in such a device. Those impellers are denoted as W13 and W17. The W17 impeller differs from the W13 impeller only by the shape of a vane. Both impellers had eight vanes each. The analysis of the impellers with the two-dimensional peculiarity method for non-viscous medium suggested higher cavitation resistance of the W13 impeller.

In the first stage, calculations were made of the undetermined flow through the pump without cavitation in order to determine the most favourable boundary conditions to be applied when analysing the flow through the pump and determining the calculation characteristics of the pump and the impeller.

The calculations reflect the full three-dimensional geometry of the pump (**Figure 11**) consisting of a straight section of the pipeline before the inlet to the impeller, a centrifugal impeller, a spiral collecting volute, a diffuser, and a short section of pipeline after the pump.

Separate discreet models have been built in the inlet and outlet impeller areas. On the cylindrical surface between the impeller and the volute, these models were not connected by common nodes and remained unfit. Thus, during the calculation it was possible to use the "sliding mesh" technique, which is used to model the rotation of the impeller in relation to the stationary casing. The discrete model is built with approximately 1.3 million tetrahedral elements in total. The elements were also compacted near the vane surface and in the area between the impeller and the collecting channel (**Figure 12**).

The mathematical model of the flow is described by the Reynolds-averaged Navier-Stokes equations (RANS). For the description of the turbulence, a two-equation k-ε model was used. The following control surfaces were used, where static pressure was monitored during the calculation:

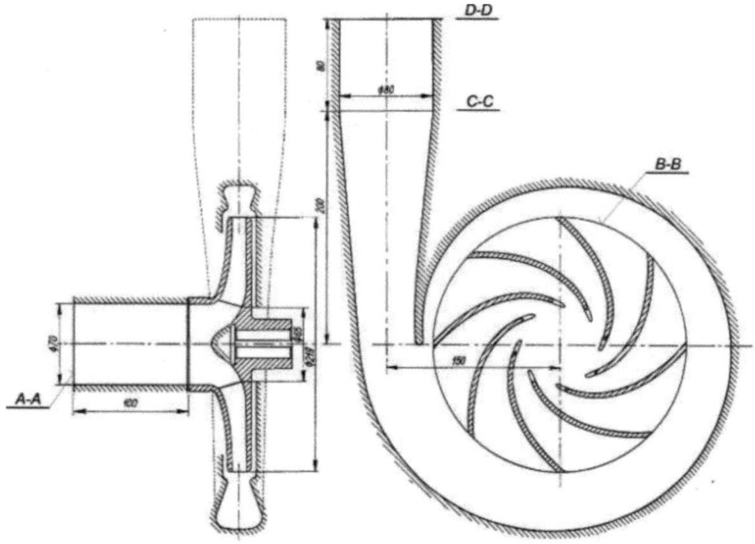

Figure 11.
The calculation area under consideration and its characteristic cross sections (W13 impeller pump).

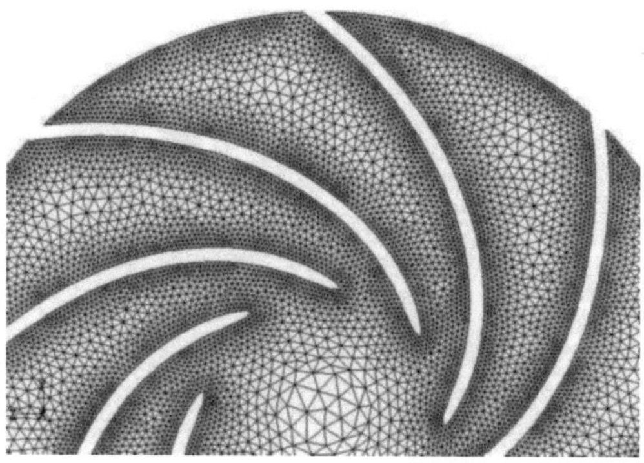

Figure 12.
Discrete model by type of tetrahedral element of the impeller surface, on the hub and rear disc side.

- The inlet section at the beginning of the suction channel (A-A)
- The cylindrical surface at the outlet from the impeller inter-vane channel (B-B)
- The cross section at the end of the diffuser (C-C)
- The outlet section at the end of the cylindrical section of the pipeline (D-D)

Calculations were made according to the scheme:

In the inlet section (A-A), a homogeneous velocity field was set with the value resulting from the flow rate and the channel section area $c = Q/A$ and the direction corresponding to the connector axis ("velocity inlet" boundary condition). In cross section (D-D), a high static pressure of 1000 kPa was set so that the pressure in the impeller would not drop below the saturation vapour pressure ("pressure outlet" boundary condition). A two-phase flow "mixture" model was selected for the calculations. During the calculations, equations describing the formation of the gaseous phase (cavitation) were excluded. This approach is suggested by ANSYS Fluent.

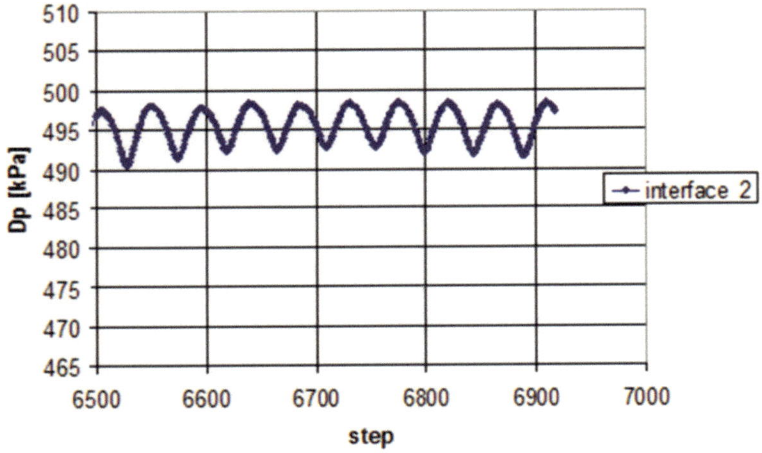

Figure 13.
Example pressure pulsation diagram as pressure difference between vane outlet (interface_2) and inlet (inlet), depending on iteration (time).

On the internal walls of the flow channel, the condition of zero velocity of the fluid in relation to the wall was set. The increase of static pressure (increase of hydrostatic height) between inlet and outlet cross sections of the pump was the expected value and allowed to reproduce flow characteristics. During the calculations, the average static pressure was monitored on the four control surfaces mentioned above. The calculations were interrupted after repeated oscillations of the static pressure on these surfaces were obtained, which took place after 6–8 rotations of the impeller. An example of a pressure pulsation diagram is shown in **Figure 13**. A fixed time step of Δt = 5,75E–5 s, corresponding to an impeller rotation by 1°, was used for the calculations.

The calculated flow characteristics of the entire pump and the W13 impeller are presented in **Figure 14**. The course of the relevant experimental characteristics is also presented.

The pump characteristics indicate a pressure increase between the cross sections A-A and C-C, characteristics of the impeller—between sections A-A and B-B. The pressure drop in the suction channel is insignificant compared to the pressure drop in the impeller.

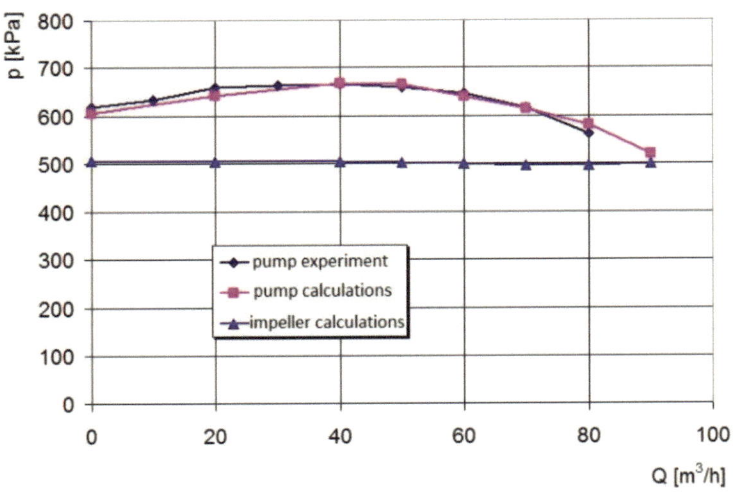

Figure 14.
Pump and impeller flow characteristics W13 determined by calculation of the transient flow (spiral collecting channel model): comparison with experimental data.

Cavitation in the pump is associated with a pressure drop in the suction area of the first degree [11]. This causes the fluid-vapour biphasic flow to occur and the continuity of the flow through the pump to be interrupted. In centrifugal pumps, cavitation shall be characterised by a clearly visible disturbance in the following characteristics: flow H = f(Q), power consumption P = f(Q) and efficiency η = f(Q). If the suction height increases at a given velocity and flow rate (or the intake height decreases), then the boundary value of the suction height at which the pump enters the cavitation state is obtained. In this way, taking into account a certain safety margin, it is possible to obtain a curve of the required excess of the energy of a fluid at the pump inlet section over the energy of evaporation of this fluid in the form of NPSH = f(Q) (net positive suction head). The NPSH parameter expresses the “suction power” of the pump:

$$NPSH = \frac{p_s - p_v}{\gamma} + \frac{c_s^2}{2g} \quad (2)$$

where p_s is the absolute pressure at the inlet cross section of the pump and c_s is the fluid velocity at the pump inlet cross section (average). Typically, this surplus is related to a state where the first-stage total head drops by 3% (NPSH3).

Determination of the cavitation state in the impeller for a given flow rate requires many calculations of the pressure distribution in the inter-vane space at the decreasing inlet pressure. The simulation assumes that a simplified geometric model of a collective channel can be used to determine the flow characteristics of the impeller itself. Instead of a spiral, an axial-symmetrical guide was used as a drainage element for the medium.

Due to the symmetry of geometry, the flow through the impeller is determined. The elimination of pressure pulsations has significantly accelerated the iterative calculation process. The flow field in the impeller still remained a periodic-symmetric field, but it was the same in all the vane channels. This allowed the calculation area to be limited to one inter-vane channel of the impeller. As a result, the calculation time corresponding to one characteristic point has been reduced.

A discrete model consisting of about 300,000 hexahedral cells was used. Since the discrete model remains stationary during the calculation, a *moving reference frame* was used which rotates at the impeller velocity.

For the calculations, the “velocity inlet” and “pressure outlet” boundary conditions were used on the outer surface of the annular collecting channel and the two-phase flow “mixture” model. During the calculation, the average static pressure value at the cross sections A-A (inlet) and B-B (outlet from the impeller inter-vane channel) was monitored.

Cavitation test in the impeller was performed for several selected values of the flow rate. Calculations were carried out in which equations describing cavitation and two-phase flow were included. The static pressure at the outlet was gradually reduced from 800 to 580 kPa.

It was found that the lowest pressure in the impeller was initially higher than the saturated vapour pressure $p_{min} > p_v$; then it was already limited by the p_v value. For each set outlet pressure, the static inlet pressure was recorded. In the W13 impeller, cavitation occurs on the impeller vanes, close to the incidence edge on the concave side of the vane. In the W17 impeller, cavitation appears on the convex side of the vane (for Q = 70 m^3/h). Selected images of the development of cavitation are presented in **Figures 15** and **16**.

When the outlet pressure is further reduced, it reaches a constant boundary value, depending on the flow rate—fully developed cavitation. Further lowering of the outlet pressure leads to a loss of convergence and interruption of the calculation.

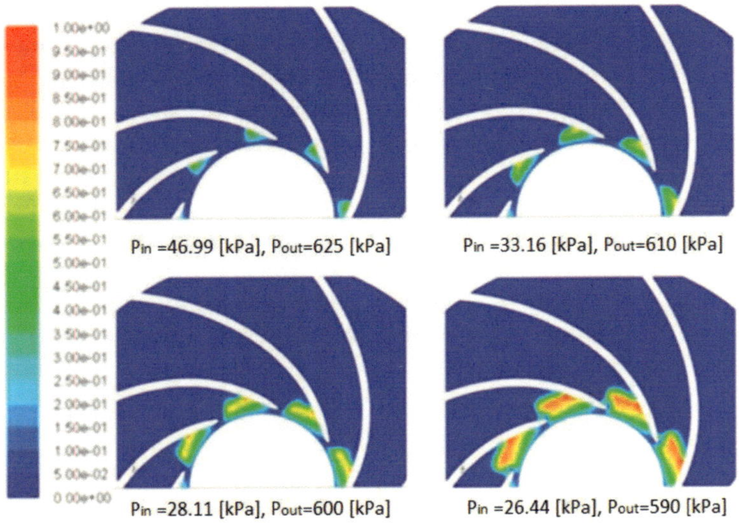

Figure 15.
Cavitation development image on the impeller W13 disc surface at Q = 70 m³/h and decreasing static pressure at the inlet (percentage of gas phase is given).

Cavitation image – model with spiral collection channel and immovable impeller (*Moving Reference Frame*) – **Figure 17**. Cavitation image – model with spiral collecting channel and rotating impeller (*Moving Mesh*) – **Figure 18**.

The cavitation fields for the axial-symmetric model are correctly symmetrical. However, the behaviour of the tested impellers is different:

- Impeller W13: cavitation is formed on the concave side of the vane.
- Impeller W17: cavitation is formed on the convex side of the vane.

The calculations converge quickly. However, the cavitation fields in the *moving reference frame* model are non-physical, and the cavitation area expands very

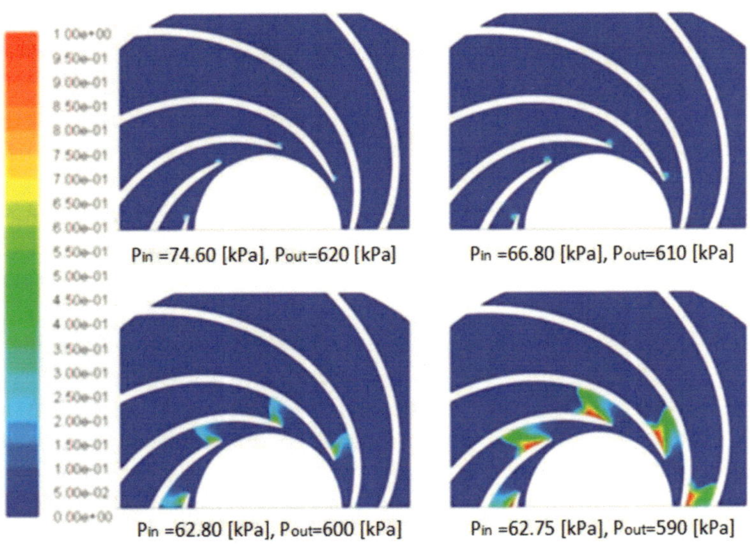

Figure 16.
Cavitation development image on the impeller W17 disc surface at Q = 70 m³/h and decreasing static inlet pressure (percentage of gas phase is given).

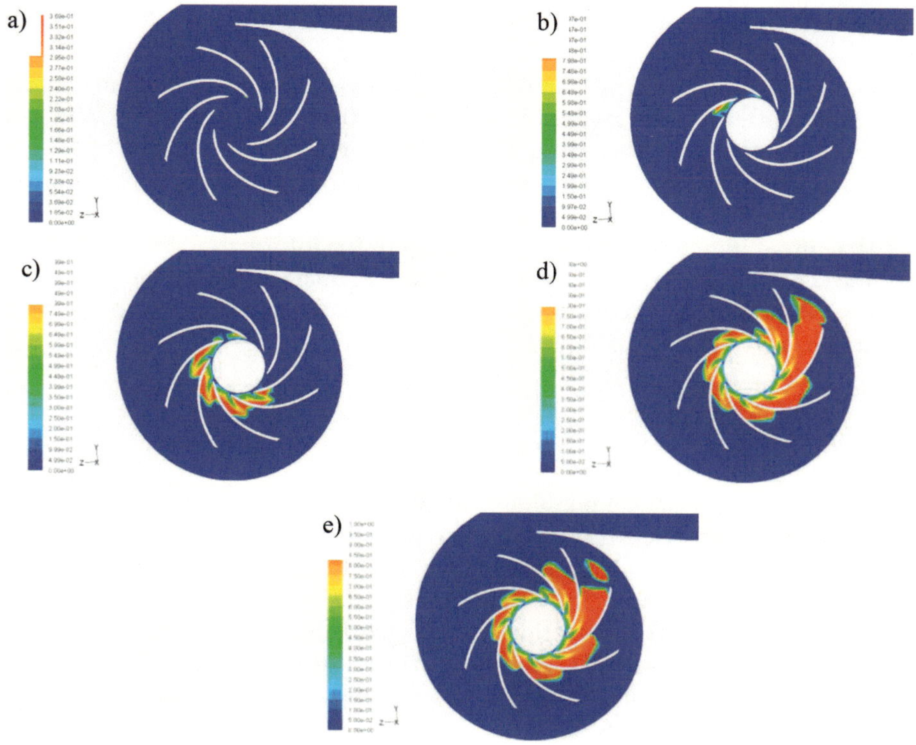

Figure 17.
Cavitation area for parameters: (a) inlet = 189 kPa and outlet = 800 kPa, (b) inlet = 95 kPa and outlet = 700 kPa, (c) inlet = 58 kPa and outlet = 650 kPa, (d) inlet = 27 kPa and outlet = 622 kPa and (e) inlet = 27 kPa and outlet = 600 kPa. The percentage of gas phase is given.

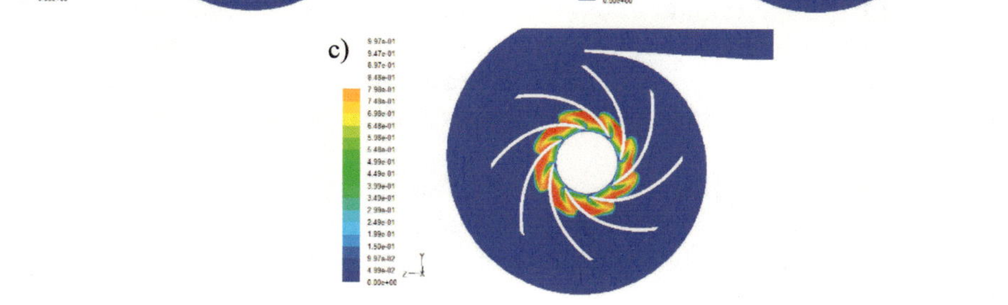

Figure 18.
Cavitation area for parameters: (a) inlet = 28.3 kPa and outlet = 675 kPa, (b) inlet = 27.8 kPa and outlet = 650 kPa and (c) inlet = 27.8 kPa and outlet = 622 kPa. The percentage of gas phase is given.

quickly. Cavitation starts in the direction of the smallest radius of the collecting spiral. The *moving mesh* model produces the best results (mainly physical). However, the problem is the slow convergence of calculations and their long duration.

5. Cavitation resistance of the pump

The different cavitation properties of the two impellers can be explained by the significantly different inlet angle of the β1 vane—30°40′ (W13) and 21° (W17)—as with the same other geometric data, resulted in a very different position of the ideal inflow point. This is confirmed by the experimental characteristics of the pumps $H = f(Q)$ and $\eta = f(Q)$ from operation.

The analyses indicated the possibility of obtaining information on cavitation resistance of the designed structure through the rational use of CFD programs. The alternative solution of designing a prototype pump and carrying out a series of experiments may be challenging.

6. Conclusion

The CFD analysis made it possible to identify areas where cavitation is more likely to occur and to assess its intensity in relation to the rotational velocity. The results showed that one of the inlet channels has both negative pressure and increased fluid flow velocity. Calculations made for different pump rotational velocities and different suction channel geometries have shown that the intensity of these phenomena increases with the rotational velocity. However, these phenomena are not strong enough to contribute to the development of the phenomenon of cavitation. A series of simulations for different suction channel geometries have confirmed that no modification of the suction channel geometry is required. Considering the designs presented here the cavitation occurred either on the convex or concave side of the vane. The main difference between the vanes was the angle of its inclination. Hence there is a specific angle between 30 and 21° at which the transition occurs. The volumetric flow rate was unchanged in both of the impeller designs, although the inlet pressures were found to be different. For blades inclined at 30°, the inlet pressures were almost twice lower than in the case of 21°. Hence lower inclination of blades is more immune to cavitation development.

The CFD calculations were made to check the selection of the main dimensions of the radial fan. After performing many variant calculations, it was found that by changing the number of blades and the outlet angle of the blades, it is possible to increase the efficiency of the fan. It appeared that the efficiency is greater for impellers with greater amount of vanes. Furthermore the efficiency increased when the vanes were inclined to 23°, and as stated above, at such angle the cavitation occurs at higher inlet pressures and represent higher immunity to cavitation. Therefore the increase of efficiency may be partially a consequence of lack of cavitation.

It was found that the characteristics of the centrifugal pump from CFD calculations are consistent with the characteristics obtained experimentally. Based on the CFD analysis, cavitation resistance of the designed centrifugal pump was determined.

Author details

Wieslaw Fiebig*, Paulina Szwemin and Maciej Zawislak
Faculty of Mechanical Engineering, Wroclaw University of Science and Technology, Wroclaw, Poland

*Address all correspondence to: wieslaw.fiebig@pwr.edu.pl

References

[1] Fiebig W, Cependa P, Jedraszczyk P, Kuczwara H. Innovative solution of an integrated motor pump assembly. In: ASME/BATH 2017 Symposium on Fluid Power and Motion Control. Sarasota, USA: ASME; 2017

[2] Frosina E. A three dimensional cfd modeling methodology applied to improve hydraulic components performance. Energy Procedia. 2015;**82**:950-956

[3] Mancò S, Nervegna N, Rundo M, Armenio G. Modelling and Simulation of Variable Displacement Vane Pumps for IC Engine Lubrication, SAE Technical Paper 2004-01-16012004

[4] Inaguma Y. Theoretical analysis of mechanical efficiency in vane pump. JTEKT Engineering Journal, English Edition No. 1007E Technical paper. 2010:28-35

[5] Frosina E, Stelson KA, et al. Vane pump power split transmission: Three dimensional computational fluid dynamics modeling. In: ASME/BATH 2015 Symposium on Fluid Power & Motion Control. Chicago, USA; 2015

[6] Houzeaux G, Codina R. A finite element method for the solution of rotary pumps. Computers and Fluids. 2007;**36**:667-679

[7] Hyun K, Hazel M, Suresh P. Two-Dimensional CFD Analysis of a Hydraulic Gear Pump. Washington: American Society for Engineeguide Education; 2007

[8] Mochala M. Intermittent CFD simulation of interlocked hydraulic pumps industrial use, basic conditions and prospect. In: FLUIDON Konferenz "Simulation in mechanischen Umfeld", Aachen; 2009

[9] Stryczek J et al. Visualisation research of the flow processes in the outlet chamber—outlet bridge—inlet chamber zone of the gear pumps. Archives of Civil and Mechanical Engineering. 2014;**15**(1):95-108. DOI: 10.1016/j.ACME 2014.02.010

[10] Jedraszczyk P, Fiebig W. CFD Model of an External Gear Pump Proceedings of 13th International Conference Computer Aided. Springer International Publishing; 2017. Available from: https://www.spguideerprofessional.de/en/cfd-model-of-an-external-gear-pump/12181546

[11] Tabaczek T, Zawiślak M, Zieliński AK. Flow and cavitation analysis of a centrifugal pump. Systems: Journal of Transdisciplinary Systems Science. 2012;**16**(2):385-394